Feynman's Tips on Physics

FEYNMAN'S
TIPS *on* PHYSICS

REFLECTIONS · ADVICE · INSIGHTS · PRACTICE

A problem-solving supplement to
The Feynman Lectures on Physics

Richard P. Feynman

Michael A. Gottlieb

Ralph Leighton

With a memoir by
Matthew Sands

BASIC BOOKS
A MEMBER OF THE PERSEUS BOOKS GROUP
New York

Copyright © 2013 by Carl Feynman, Michelle Feynman, Michael A. Gottlieb, Ralph Leighton
Published by Basic Books,
A Member of the Perseus Books Group

Books published by Basic Books are available at special discounts for bulk purchases in the United States by corporations, institutions, and other organizations. For more information, please contact the Special Markets Department at the Perseus Books Group, 2300 Chestnut Street, Suite 200, Philadelphia, PA 19103, or call (800) 810-4145, ext. 5000, or e-mail special.markets@perseusbooks.com.

A CIP catalog record for this book is available from the Library of Congress.
LCCN: 2011944291

ISBN: 978-0-465-02797-2 (paperback)
ISBN: 978-0-465-02921-1 (e-book)

10 9 8 7 6 5 4 3 2 1

Contents

Preface to the Second Edition

In the six years since the initial publication of *Feynman's Tips on Physics* (Addison-Wesley, 2006) interest in this supplement to *The Feynman Lectures on Physics* has continued unabated, as evidenced by the ever-increasing number of visitors to The Feynman Lectures Website (`www.feynman lectures.info`), created in conjunction with this project: thousands of inquiries have come in, many of them reporting suspected errata in *The Feynman Lectures*, and many with questions and comments about physics exercises.

It is thus with great pleasure and pride we present this second edition of *Feynman's Tips on Physics*, published by Basic Books as part of a unification of print, audio, and photo rights pertaining to *The Feynman Lectures on Physics*—rights which had been assigned over the years to different publishers. To celebrate this fortuitous occasion, *The Feynman Lectures on Physics* (*New Millennium Edition*) is now being printed for the first time from a LaTeX manuscript, thus enabling errata to be corrected much more quickly, and electronic editions of *The Lectures* to be produced soon. In addition, this new edition of *Feynman's Tips on Physics* is being made available in softcover at a greatly reduced price from the hardcover original, and expanded to include three insightful interviews about *The Lectures:*

- with Richard Feynman, in 1966, soon after his key part in the project was finished,

- with Robert Leighton, in 1986, about Feynman's gifts as a lecturer— and the challenges of translating from "Feynmanese" into English, and

- with Rochus Vogt, in 2009, about the community of professors that cooperatively taught *The Feynman Lectures* course at Caltech.

To all of you who e-mailed or posted questions and comments about *The Feynman Lectures on Physics* and *Feynman's Tips on Physics,* we wish to offer our heartfelt thanks; your contributions and support have helped greatly to improve these books, and will be appreciated by future generations of readers. To those who wrote requesting more exercises, we apologize that they could not be included in this edition. However, your encouragement has inspired the creation of a new, expansive (soon-to-be-published) book, *Exercises for The Feynman Lectures on Physics.*

Michael A. Gottlieb
Ralph Leighton
November 2012

Foreword

At a lonely border post high on the Himalayan frontier, Ramaswamy Balasubramanian peered through his binoculars at the People's Liberation Army soldiers stationed in Tibet—who were peering through their scopes back at him. Tensions between India and China had been high for several years since 1962, when the two countries traded shots across their disputed border. The PLA soldiers, knowing they were being watched, taunted Balasubramanian and his fellow Indian soldiers by shaking, defiantly, high in the air, their pocket-sized, bright-red copies of *Quotations from Chairman Mao*—better known in the West as "Mao's Little Red Book."

Balasubramanian, then a conscript studying physics in his spare time, soon grew tired of these taunts. So one day, he came to his observation post prepared with a suitable rejoinder. As soon as the PLA soldiers started waving Mao's Little Red Book in the air again, he and two fellow Indian soldiers picked up and held aloft the three big, bright-red volumes of *The Feynman Lectures on Physics*.

One day I received a letter from Mr. Balasubramanian. His was among hundreds of letters I have received over the years that describe the lasting impact Richard Feynman has had on people's lives. After recounting the "red-books" incident on the Sino-Indian frontier, he wrote: "Now, twenty years later, whose red books are still being read?"

Indeed. Today, more than forty years after they were delivered, *The Feynman Lectures on Physics* are still being read—and still inspire—even in Tibet, I suspect.

A special case in point: several years ago I met Michael Gottlieb at a party where the host was displaying on a computer screen the harmonic overtones of a live Tuvan throat-singer—the kind of event that makes living in San Francisco such fun. Gottlieb had studied math and was very interested in physics, so I suggested he read *The Feynman Lectures on Physics*—and about a year later, he devoted six months of his life to reading *The Lectures* very carefully from beginning to end. As Gottlieb describes in his introduction, this led, eventually, to the book you are reading now, as well as to a new, "Definitive Edition" of *The Feynman Lectures on Physics*.

Thus I am pleased that people interested in physics all over the world can now study, with the addition of this supplemental volume, a more correct and complete edition of *The Feynman Lectures on Physics*—a monumental work that will continue to inform and inspire students for decades to come, whether in midtown Manhattan or high in the Himalayas.

Ralph Leighton
May 11, 2005

Richard Feynman, circa 1962

Introduction

I first heard of Richard Feynman and Ralph Leighton in 1986, through their entertaining book *Surely You're Joking, Mr. Feynman!* Thirteen years later I met Ralph at a party. We became friends, and over the next year we worked together on the design of a fantasy stamp honoring Feynman.[1] All the while Ralph was giving me books to read, by or about Feynman, including (since I am a computer programmer) *Feynman Lectures on Computation.*[2] The discussion of quantum mechanical computation in this fascinating book intrigued me, but without having studied quantum mechanics, I had difficulty following the arguments. Ralph recommended I read *The Feynman Lectures on Physics Volume III: Quantum Mechanics,* which I began, but Chapters 1 and 2 of *Volume III* are reproduced from Chapters 37 and 38 of *Volume I,* so I found myself backtracking through references in *Volume I* rather than progressing through *Volume III.* I therefore decided to read all *The Feynman Lectures* from beginning to end—I was determined to learn some quantum mechanics! However, that goal became secondary as time

[1]Our stamp appears in the liner notes of *Back TUVA Future,* a CD featuring the Tuvan throat-singing master *Ondar* and a cameo appearance by Richard Feynman (Warner Bros. 9 47131-2), released in 1999.

[2]*Feynman Lectures on Computation,* by Richard P. Feynman, edited by Anthony J.G. Hey and Robin W. Allen, 1996, Addison-Wesley, ISBN 0-201-48991-0.

went on and I became increasingly absorbed in Feynman's fascinating world. The joy of learning physics, simply for the pleasure of it, became my highest priority. I was hooked! About halfway through *Volume I*, I took a break from programming and spent six months in rural Costa Rica studying *The Lectures* full-time.

Every afternoon I studied a new lecture and worked on physics problems; in the mornings I reviewed and proofread yesterday's lecture. I was in e-mail contact with Ralph, and he encouraged me to keep track of errors I mentioned encountering in *Volume I*. It was not much of a burden, because there were very few errors in that volume. However, as I progressed through *Volumes II* and *III*, I was dismayed to discover increasingly more errors. In the end I had compiled a total of more than 170 errors in *The Lectures*. Ralph and I were surprised: how could so many errors have been overlooked for so long? We decided to see what could be done about getting them corrected in the next edition.

Then I noticed some intriguing sentences in Feynman's preface:

"The reason there are no lectures on how to solve problems is because there were recitation sections. Although I did put in three lectures in the first year on how to solve problems, they are not included here. Also there was a lecture on inertial guidance which certainly belongs after the lecture on rotating systems, but which was, unfortunately, omitted."

This suggested the idea of reconstructing the missing lectures and, if they proved interesting, offering them to Caltech and Addison-Wesley for inclusion in a more complete and error-corrected edition of *The Lectures*. But first I had to *find* the missing lectures, and I was still in Costa Rica! Through a bit of deductive logic and investigation, Ralph was able to locate the lecture notes, which were previously hidden away somewhere between his father's office and the Caltech Archives. Ralph also obtained tape recordings of the missing lectures, and while researching errata in the Archives after my return to California, I fortuitously discovered the blackboard photos (long believed lost) in a box of miscellaneous negatives. The Feynman heirs generously gave us permission to use these materials, and so, with some useful critiques from Matt Sands, now the only surviving member of the Feynman-Leighton-Sands trio, Ralph and I reconstructed *Review B* as a sample, and presented it with the errata for *The Lectures* to Caltech and Addison-Wesley.

Addison-Wesley received our ideas enthusiastically, but Caltech was initially skeptical. Ralph therefore appealed to Kip Thorne, the Richard Feynman Professor of Theoretical Physics at Caltech, who eventually managed to achieve a mutual understanding among all involved, and who generously volunteered his time to oversee our work. Since Caltech did not want to amend the existing volumes of *The Lectures* for historical reasons, Ralph proposed putting the missing lectures in a separate book. That is the origin of this supplementary volume. It is being published in parallel with a new *Definitive Edition* of *The Feynman Lectures on Physics,* in which the errors I found are corrected, as are other errors found by a number of other readers.

Matt Sands' memoir

In our quest to reconstruct the four lectures, Ralph and I had many questions. We felt very fortunate to be able to get answers from Professor Matt Sands, the man whose idea it was to embark on the ambitious project that produced *The Feynman Lectures on Physics.* We were surprised that the story of their genesis was not widely known, and realizing that this project offered an opportunity to remedy that deficit, Professor Sands kindly agreed to write a memoir on the origins of *The Feynman Lectures* for inclusion in this supplement.

The four lectures

From Matt Sands we learned that in December 1961, toward the end of the first term[3] of Feynman's Caltech freshman physics course, it was decided that it would be unfair to introduce new material to the students just a few days before the final exam. So, for the week preceding the test, Feynman gave three optional review lectures, in which no new material was introduced. The review lectures were intended for students having difficulties in the class, and emphasized techniques for understanding and solving physics problems. Some of the example problems were of historical interest, including the discovery of the atomic nucleus by Rutherford, and the determination of the mass of the pi meson. With characteristic human insight, Feynman also discussed the solution to another kind of problem, equally important to at least half the students in his freshman class: the emotional problem of finding oneself below average.

The fourth lecture, *Dynamical Effects and Their Applications,* was given early in the second term of the freshman class, shortly after the students returned from winter break. Originally, it was to be *Lecture 21,* and the idea behind it was to take a rest from the difficult theoretical discussion of

---

[3]The academic year at Caltech is divided into three terms; the first runs from late September to early December, the second from early January to early March, and the third from late March to early June.

rotations presented in Chapters 18 through 20 and show the students some interesting applications and phenomena that arise from rotations, "just for entertainment." Most of the lecture was devoted to a discussion of technology that was relatively new in 1962: practical inertial guidance. The remainder of the lecture discussed natural phenomena that arise from rotations, and also offered a clue as to why Feynman described the omission of this lecture from *The Feynman Lectures on Physics* as "unfortunate."

After the lecture

After ending a lecture Feynman often left his microphone on. This has provided us with the unique opportunity of witnessing how Feynman interacted with his undergraduate students. The example given here, recorded after *Dynamical Effects and Their Applications,* is particularly noteworthy for its discussion of the incipient transition in real-time computing from analog to digital methods in 1962.

The exercises

In the course of this project Ralph reestablished contact with his father's good friend and colleague Rochus Vogt, who graciously gave his permission to republish exercises and solutions from *Exercises in Introductory Physics,* the collection that Robert Leighton and he had created especially for *The Lectures* back in the 1960s. Due to space limitations I chose only exercises for *Volume I,* Chapters 1 through 20 (the material covered before *Dynamical Effects and Their Applications*), preferring problems that, to quote Robert Leighton, "are numerically or analytically simple, yet incisive and illuminating in content."

Website

Readers are invited to visit www.feynmanlectures.info for more information on this volume and *The Feynman Lectures on Physics.*

Michael A. Gottlieb
Playa Tamarindo, Costa Rica
mg@feynmanlectures.info

Acknowledgments

We wish to express our heartfelt thanks to all who made this book possible, especially:

Thomas Tombrello, Chairman of the Physics, Mathematics, and Astronomy Division, for approving this project on behalf of Caltech;

Carl Feynman and Michelle Feynman, heirs to Richard Feynman, for their permission to publish their father's lectures in this book;

Marge L. Leighton, for her permission to publish an excerpt from the *Oral History of Robert B. Leighton,* and problems from *Exercises in Introductory Physics;*

Matthew Sands, for his wisdom, knowledge, constructive comments, and suggestions on the manuscript—and for his illuminating memoir;

Rochus E. Vogt, for the ingenious problems and answers in *Exercises in Introductory Physics,* for his interview with us, and for his permission to use them in this volume;

Michael Hartl, for his meticulous proofreading of the manuscript, and for his diligent work with the errata in *The Feynman Lectures on Physics;*

John Neer, for studiously documenting Feynman's lectures at the Hughes Aircraft Corporation, and for sharing those notes with us;

Helen Tuck, Feynman's secretary for many years, for her encouragement and support;

Adam Cochran, for his consummate skill in navigating the swamp of tangled book contracts and personalities to find a new home for this book, as well as for *The Feynman Lectures on Physics;* and

Kip Thorne, for his grace and tireless work securing the trust and support of everyone involved, and for overseeing our work.

On the Origins of
The Feynman Lectures on Physics

A MEMOIR BY MATTHEW SANDS

Education reform in the 1950s

When I first became a regular faculty member at Caltech in 1953, I was asked to teach some graduate courses. I found myself quite dismayed about the course program for the graduate students. During the first year they were given courses only in classical physics—mechanics and electricity and magnetism. (And even the E and M class covered only statics, no radiation theory at all.) I thought it was disgraceful that these hotshot students were not exposed to the ideas of modern physics (many of which had already been around for 20 to 50 or more years) until their second or third year in grad school. So I began a campaign to reform the program. I had known Richard Feynman since our days at Los Alamos, and we had both come to Caltech a few years back. I asked Feynman to join the campaign, and we outlined a new program and eventually persuaded the physics faculty to adopt it. The first year program consisted of a course in Electrodynamics and Electron Theory (taught by me), Introductory Quantum Mechanics (taught by Feynman), and, as I recall, a course in Mathematical Methods, taught by Robert Walker. I think that the new program was quite successful.

At about that time Jerrold Zacharias of MIT was stimulated by the appearance of *Sputnik* to push for a program to revitalize the teaching of high school physics in the United States. One result was the creation of the PSSC (Physical Science Study Committee) program, and the generation of many new materials and ideas, as well as some controversy.

When the PSSC program was nearing its completion, Zacharias and some colleagues (I believe among them Francis Friedman and Philip Morrison) decided that it was time to tackle also a revision of university physics. They organized a couple of large meetings of physics instructors, out of which came the formation of the Commission on College Physics, a national committee of a dozen university physics instructors, which was supported by the National Science Foundation, and was charged with stimulating some national endeavors for the modernization of physics teaching in colleges and universities. Zacharias invited me to those first meetings and I later served on the Commission, eventually becoming its chairman.

The Caltech program

These activities prompted me to begin thinking about what could be done with the Caltech undergraduate program, with which I had long been rather unhappy. The introductory course in physics was based on the book of Millikan, Roller, and Watson, a very fine book that had been written, I believe, in the 1930s, and, though revised later by Roller, had little or no modern physics. Further, the course was taught without lectures, so there was little opportunity to introduce new material. The strength of the course was a set of intricate "problems" compiled by Foster Strong[1], which were used for weekly homework assignments, and two weekly recitation sections in which the students discussed the assigned problems.

Like other physics faculty, I was each year assigned to be the advisor to a handful of physics majors. When talking with the students, I was often dismayed that by their junior year these students were getting discouraged about continuing in physics—it seemed at least in part because they had been studying physics for two years, but still had not been exposed to any of the ideas of current physics. So it was that I decided not to wait for the national program to mature, but to try to do something at Caltech. In particular, I wanted to see some of the content of "modern" physics—atoms, nuclei, quanta, and relativity—brought into the introductory course. After discussions with a few colleagues—most notably Thomas Lauritsen and Feynman—I proposed to Robert Bacher, then head of physics, that we should start a program to reform the introductory course. His initial response was not very encouraging. He said, in effect: "I have been telling people we have a very fine program that I am proud of. Our discussion sections are staffed by some of our senior faculty. Why should we change?" I persisted and was supported by a few others, so Bacher relented, accepted the idea, and had soon secured a grant from the Ford Foundation (for, if I remember correctly, something more than a million dollars). The grant was to be used for the costs of devising new equipment for the introductory

---

[1]The exercises in Chapter 5 of this volume include more than a dozen problems from Foster Strong's collection that were reproduced with permission in *Exercises in Introductory Physics* by Robert B. Leighton and Rochus E. Vogt.

labs, and for developing new content for the course—in particular, for some temporary faculty to pick up the regular duties of the ones who were devoting time to the project.

When the grant was received, Bacher appointed a small task force to lead the program: Robert Leighton, as chairman, Victor Neher, and me. Leighton had long been involved in the upper division program—of which his book *Principles of Modern Physics*[2] was the mainstay—and Neher was known as a brilliant instrumentalist. I was, at the time, miffed that Bacher had not asked me to be the leader of the group. I guessed that it may have been partly because I was already fairly busy running the Synchrotron Laboratory, but I have always thought that he was also worried that I might be too "radical" and that he wanted to balance the project with Leighton's conservatism.

The committee agreed from the start that Neher would concentrate on developing new labs—about which he had many ideas—and that we should work toward presenting a lecture course in the following year—feeling that the lectures would provide the best mechanism for developing a new course content. Leighton and I were to design a syllabus for the lectures. We began by working independently to produce course outlines, but meeting weekly to compare progress and to try to reach a common ground.

Impasse and inspiration

It soon became clear that a common ground was not easily to be found. I usually saw Leighton's approach to be too much of a rehash of the content of physics courses that had been in vogue for 60 years. Leighton thought that I was pushing impractical ideas—that freshmen were not ready for the "modern" content I wanted to introduce. I was, fortunately, bolstered in my resolve by frequent conversations with Feynman. Feynman was already well known as an impressive lecturer, and was particularly adept at explaining the ideas of modern physics to a general audience. I would frequently stop at his house on the way home from the Institute to sound him out on what I was thinking, and he would often make suggestions about what might be done, and was generally supportive.

After several months of these efforts, I became rather discouraged; I didn't see how Leighton and I could ever come to an agreement on a syllabus. Our concepts for the course seemed to be completely at odds. Then

---

[2]*Principles of Modern Physics,* by Robert B. Leighton, 1959, McGraw-Hill, *Library of Congress Catalog Card Number* 58-8847.

one day I had an inspiration: Why not ask Feynman to give the lectures for the course? We could provide him with the outlines of both Leighton and myself, and let him decide what to do. I immediately proposed this idea to Feynman in the following way: "Look, Dick, you have now spent forty years of your life seeking an understanding of the physical world. Here is an opportunity for you to put it all together and present it to a new generation of scientists. Why don't you give the freshman lectures next year?" He was not immediately enthusiastic, but we continued over the next few weeks to discuss the idea, and he was soon caught up in the notion. He would say maybe we could do this or that. Or this would fit in here, and so on. After a few weeks of such discussions, he asked me: "Has there ever been a great physicist who has presented a course to freshmen?" I told him that I didn't think that there ever had been. His response: "I'll do it."

Feynman will give the lectures

At the next meeting of our committee I presented with great enthusiasm my proposal—only to be dismayed by the cool response of Leighton. "That's not a good idea. Feynman has never taught an undergraduate course. He wouldn't know how to speak to freshmen, or what they could learn." But Neher saved the day. His eyes lit up with excitement and he said: "That would be great. Dick knows so much physics, and knows how to make it interesting. It would be fantastic if he would really do it." Leighton was persuaded, and once persuaded, supported the idea wholeheartedly.

Some days later I faced the next hurdle. I presented the idea to Bacher. He didn't think much of it. He considered that Feynman was too important to the graduate program and could not be spared. Who would teach quantum electrodynamics? Who would be working with the theoretical graduate students? And besides, could he really bend down to the level of the freshmen? At this point I did some lobbying with some of the senior members of the physics department, who put in some supporting words to Bacher. And finally, I used the argument dear to academics: If Feynman really wants to do it, do you want to say that he should not? The decision was made.

With six months remaining before the first lecture, Leighton and I talked with Feynman about what we had been thinking. He started to work intensively on developing his own ideas. At least once each week I would stop by his house, and we would discuss what he had been thinking. He would sometimes ask whether I thought that some particular approach would be accessible to the students, or whether I would think that this or that sequence of material would "work" best. I may mention a particular example. Feynman had been working on how to present the ideas of wave interference and diffraction, and was having difficulty finding a suitable mathematical approach—one both straightforward as well as powerful. He was not able to come up with one without the use of complex numbers. He asked me whether I thought the freshmen would be able to work with the algebra of complex numbers. I reminded him that the students admitted to Caltech had been selected primarily on their demonstrated abilities with mathematics, and that I was confident that they would not have problems dealing with complex algebra, so long as they were given some brief introduction to the subject. His twenty-second lecture contains a delightful introduction to the algebra of complex quantities, which he was then able to use in many of the following lectures for the description of oscillating systems, for problems in physical optics, and so on.

Early on, a small problem surfaced. Feynman had a long-time commitment to be absent from Caltech for the third week of the fall term, and so would miss two class lectures. We agreed that that problem was easily solved. I would substitute for him on those days. However, in order not to break the continuity of his presentation, I would give the two lectures on some subsidiary topics that might be useful to the students, but were not related to his main line of development. This explains why Chapters 5 and 6 of *Volume I* are somewhat anomalous.

For the most part, however, Feynman worked alone at developing a complete outline of what he would do for the whole year—filling in enough detail to be sure that there would not be unforeseen difficulties. He worked intensely for the rest of that academic year, and by September (now 1961) was ready to begin his first year of lectures.

The new physics course

Originally it was considered that the lectures given by Feynman would form the starting point of an evolution of a revised program for the two-year introductory course—one required of all of the incoming students at Caltech. It was thought that in succeeding years others of the faculty would take over responsibility for each of the two years, developing eventually a "course"—with a textbook, homework exercises, a laboratory, and so on.

For the first years of the lectures, however, a different format needed to be devised. No course materials were available and had to be created as we went along. Two one-hour lectures were scheduled—at 11 a.m. on Tuesday and Thursday, and students were assigned to a one-hour discussion section each week, which was led by a faculty member or a graduate student assistant. There was also a three-hour laboratory each week, under the direction of Neher.

During the lectures Feynman carried a microphone, suspended from his neck and coupled to a magnetic tape recorder in another room. Photographs were periodically taken of the contents of the blackboards. Both services were managed by Tom Harvey, the technical assistant in charge of the lecture hall. Harvey also helped Feynman devise an occasional demonstration for the lectures. The recorded lectures were transcribed to a rather legible form by a typist, Julie Cursio.

That first year Leighton took on the responsibility for seeing that the transcripts were edited for clarity, and as quickly as possible, so that the students would have the printed lecture notes for study soon after the lectures were given. It was thought at first that this job could be done by assigning each lecture to one of the graduate students who were leading the discussion sections and labs. That didn't work out, however, because it was taking the students much too long, and the resulting product reflected more the ideas of the student than those of Feynman. Leighton quickly changed the arrangement by taking on much of the work himself, and by recruiting various faculty members (from physics and engineering) to take on the job of editing one or more of the lectures. Under this plan, I also edited several of the lectures during that first year.

For the second year of the course some changes were made. Leighton took over the responsibility for the first-year students—giving the lectures and generally managing the course. Fortunately, the students now had available from the beginning the transcribed notes of Feynman's lectures from the previous year. I became responsible for looking after the details of the second-year course, for which Feynman was now giving the lectures. And I was left with the responsibility of producing in a timely manner the edited transcripts. Because of the nature of the second-year material, I concluded that it would be most appropriate to take on the task myself.

I also sat in on nearly all of the lectures—as I had done during the first year—and took one of the discussion sections for myself, so that I could see how the course was going for the students. After each lecture, Feynman, Gerry Neugebauer, and I, occasionally with one or two others, would usually go to lunch at the student cafeteria, where we would have a discussion about what might be suitable homework exercises to be given to the students on the subject of the lecture. Feynman would generally have in mind several ideas for these exercises, and others would emerge from the discussion. Neugebauer was responsible for collecting these exercises and producing a "problem set" each week.

What the lectures were like

It was a great pleasure to sit in on the lectures. Feynman would appear five minutes or so before the scheduled start of the lecture. He would take out of his shirt pocket one or two small pieces of paper—perhaps 5 by 9 inches—unfold them, and smooth them out at the center of the lecture bench at the front of the lecture hall. These were his notes for his lecture, though he rarely referred to them. (A photo reproduced at the beginning of Chapter 19 of *Volume II* shows Feynman during one of his lectures, standing behind the lecture bench, with two sheets of notes visible on the bench.) As soon as the bell would ring, announcing the start of the official class period, he would start his lecture. Each lecture was a carefully scripted, dramatic production, which he had, clearly, planned in detail—usually with an introduction, development, climax, and denouement. And his timing was most impressive. Only very rarely would he finish more than a fraction of a minute before or after the end of the hour. Even the use of the chalk boards at the front of the lecture hall appeared to be carefully choreographed. He would begin at the upper left of board number one on the left, and at the end of the lecture would have just completely filled board two on the far right.

But the greatest pleasure was, of course, watching the development of the original sequence of ideas—presented with clarity and style.

The decision to make a book

Although we had not initially contemplated that the lecture transcripts would become a book, that idea came into serious consideration at about the middle of the second year of the lectures—in the spring of 1963. The thought was stimulated in part by inquiries from physicists from other schools about whether transcripts could be made available to them, and in part by suggestions from several book editors—who had, of course, got wind that the lectures were going on, and had perhaps seen copies of the transcripts—that we should consider a book and that they would like to publish it.

After some discussion we decided that the transcripts could, with some work, be turned into a book, so we asked the interested publishers to make proposals to us for doing that. The most attractive proposal came from representatives of the Addison-Wesley Publishing Company (A-W), who proposed that they could provide us with hardbound books in time for the class of September, 1963—only six months after the decision to publish. Also, in view of the fact that we were not asking that the authors receive royalties, they proposed that the books could be available at a rather low price.

Such a rapid publication schedule would be possible because they had complete facilities and staff in-house for editing, and typesetting, through to photo-offset printing. And by adopting a novel (at that time) format consisting of a single wide column of text together with a very wide "margin" on one side, they could accommodate figures and other ancillary material. This format meant that what would normally be galley proofs could be used directly for the final page layouts, without any need to reset textual material to accommodate figures and the like.

The A-W proposal won the day. I took on the task of making any necessary revisions and annotations in the lecture transcripts, and generally working with the publisher—proofreading the typeset material, and so on. (Leighton was at this time heavily involved in teaching the second round of the freshman course.) I would revise each lecture transcript for clarity and accuracy, then give it to Feynman for a final check, and as soon as a few lectures were ready, would send them off to A-W.

I rather quickly sent off the first few lectures, and very soon received back the galleys for proofreading. It was a disaster! The editor at A-W had done a significant rewrite, converting the informal style of the transcripts to a traditional, formal, text-book style—changing "you" to "one", and so on. Fearing a possible confrontation on the matter, I telephoned the editor. After explaining that we considered that the informal, conversational style was an essential part of the lectures, and that we preferred personal pronouns to the impersonal ones, and so on, she saw the light and thereafter did a great job—mostly leaving things as they were. (It was then a pleasure to work with her, and I wish I could remember her name.)

The next stumbling block was more serious: choosing a name for the book. I recall visiting Feynman in his office one day to discuss the subject. I proposed that we adopt a simple name like "*Physics*" or "*Physics One*" and that the authors should be Feynman, Leighton, and Sands. He didn't particularly like the suggested title, and had a rather violent reaction to the proposed authors: "Why should your names be there—you were only doing the work of a stenographer!" I disagreed, pointing out that without the efforts of Leighton and me, the lectures would never have come to be a book. The disagreement was not immediately resolved. I returned to the discussion some days later, and together we came up with a compromise: *"The Feynman Lectures on Physics* by Feynman, Leighton, and Sands."

The Feynman preface

After the completion of the second year of lectures—near the beginning of June, 1963—I was in my office assigning the grades for the final examinations, when Feynman dropped in to say goodbye before leaving town (perhaps to go to Brazil). He asked how the students had performed on the exam. I said I thought pretty well. He asked what was the average grade, and I told him—something like 65 percent as I recall. His response was, "Oh, that's terrible, they should have done better than that. I am a failure." I tried to dissuade him of this idea, pointing out that the average grade was very arbitrary, depending on many factors such as the difficulty of the problems given, the grading method used, and such—and that we usually tried to make the average sufficiently low that there would be some spread in grades to provide a reasonable "curve" for the assignment of letter grades. (This is an attitude, incidentally, that I wouldn't approve of today.) I said that I thought that many of the students had clearly got a great deal out of the class. He was not persuaded.

I then told him that the publication of *The Lectures* was proceeding apace and wondered whether he would like to provide some kind of preface. The idea was interesting to him, but he was short of time. I suggested that I could turn on the dictating machine I had on my desk, and that he could dictate his preface. So, still thinking about his depression over the average grade on the final exam of the second year students, he dictated the first draft of *Feynman's Preface,* which you will find in front of each volume of *The Lectures.* In it he says: "I don't think I did very well by the students." I have often regretted that I had arranged for him to make a preface in this way, because I do not think that this was a very considered judgment. And I fear that it has been used by many teachers as an excuse for not trying out *The Lectures* with their students.

The second and third volumes

The story of the publication of the second year of lectures is a little different from the first year. First, when the second year came to an end (now about June of 1963) it was decided to split the lecture notes into two parts, to make two separate volumes: *Electricity and Magnetism,* and *Quantum*

Physics. Second, it was thought that the lecture notes on quantum physics could be greatly improved with some augmentation and rather extensive reworking. To this end Feynman proposed that he would, toward the end of the following year, give a number of additional lectures on quantum physics, which could be blended with the original set to make up the third volume of the printed lectures.

There was an additional complication. The federal government had a year or so earlier authorized the construction at Stanford University of a two-mile-long linear accelerator to produce 20-GeV electrons for particle physics research. It was to be the largest and most expensive accelerator yet built, with electron energies and intensities many times higher than any existing facility—an exciting project. For more than a year W.K.H. Panofsky, who had been appointed Director of the newly created laboratory—the Stanford Linear Accelerator Center—had been trying to persuade me to join him as Deputy Director, helping to build the new accelerator. In the spring of that year he prevailed, and I agreed to move to Stanford at the beginning of July. I was, however, committed to seeing *The Lectures* through to completion, so part of the arrangement was that I would take that work with me. Once at Stanford I found my new responsibilities more demanding that I had expected, so that I found it necessary to work on *The Lectures* most evenings if I was to make suitable progress. I managed to complete the final editing of *Volume II* by March of 1964. Fortunately, I had the very capable assistance of my new secretary, Patricia Preuss.

By May of that year Feynman had given the additional lectures on quantum physics, and we began to work on *Volume III.* Because some major restructuring and revision was required, I went several times to Pasadena for long consultations with Feynman. Problems were easily overcome and the material for the third volume was completed by December.

The student response

From the contact with the students in my discussion section, I could have a pretty clear impression about how they were reacting to the lectures. I believe that many, if not most, of them realized that they were having a privileged experience. I also saw that they were often caught up in the excitement of the ideas and learning a lot of physics. That did not apply, of course, to all of the students. Remember that the course was required of all incoming students, though less than one-half were planning to be physics majors, and so many of the others formed, in effect, a captive audience. Also, some of the shortcomings of the course became evident. As an example, the students often had difficulty separating the key ideas in the lectures from some of the secondary material introduced to provide illustrative applications. They found this particularly frustrating when studying for examinations.

In a special preface to the Commemorative Issue of *The Feynman Lectures on Physics,* David Goodstein and Gerry Neugebauer have written that " . . . as the course wore on, attendance by the registered students started dropping alarmingly." I don't know where they got this information. And I wonder what evidence they have that: "Many of the students dreaded the class . . . " Goodstein was not at Caltech at that time. Neugebauer was part of the crew working on the course, and would sometimes jokingly say that there were no undergraduate students left in the lecture hall—only grad students. That may have colored his memory. I was sitting at the back of the hall at most of the lectures, and my memory—of course, dimmed by the years—is that perhaps 20 percent or so of the students were not bothering to attend. Such a number would not be unusual for a large lecture class, and I do not remember that anyone was "alarmed." And although there may have been some students in my recitation section who dreaded the class, most were involved and excited by the lectures—although some of them, very likely, would have dreaded the homework assignments.

I would like to give three illustrations of the kind of impact that the lectures made on the students of those first two years. The first dates from the time that the course was being given, and though that is more than 40 years ago, it made such an impression on me that I remember it clearly. It was at the very beginning of the second year, and, by an accident of scheduling, my discussion section first met just before the first of Feynman's lectures for that year. Inasmuch as we did not have a lecture to discuss, and no homework had yet been assigned, it was not clear what we should talk about. I began the class by asking the students to tell their impressions of the previous year's lectures—which had finished some three months earlier. After some responses, one student said that he had been intrigued by the discussion of the structure of the eye of the bee, and about how it had been optimized by a balance between the effects of geometrical optics and the limitations from the wave nature of light (see *The Feynman Lectures on Physics (FLP)*, Vol. I Section 36-4). I asked whether he could reconstruct the arguments. He went to the blackboard and with very little prompting from me was able to reproduce the essential elements of the argument. And this some six months after the lecture, and with no review.

The second illustration is provided by a letter I received in 1997—some 34 years after the lectures were given—from a student, Bill Satterthwaite, who attended the lectures, as well as my recitation section. The letter came out of the blue, prompted by his encounter with an old friend of mine at MIT. He wrote:

> "This letter is to thank you and everyone else for Feynman physics. . . . Dr. Feynman's introduction says he does not think he served the students very well. . . . I disagree. I and my friends always enjoyed them and realized what a unique and wonderful experience they were. And we learned a lot. As for objective evidence about how we felt, I don't remember any other regular lecture in my Caltech career getting applause and my memory says it happened fairly often at the end of Dr. Feynman's lectures. . . . "

The last illustration dates from a few weeks ago. I happened to be reading the autobiographical sketch written by Douglas Osheroff, who was awarded the Nobel Prize in physics for 1996 (together with David Lee and Robert Richardson) for the discovery of the superfluid state in Helium-3. Osheroff wrote:

> "It was a good time to be at Caltech, as Feynman was teaching his famous undergraduate course. This two-year sequence was an extremely important part of my education. Although I cannot say that I understood it all, I think it contributed most to the development of my physical intuition."

Afterthoughts

My rather brusque departure from Caltech immediately after the second year of the lectures meant that I had no opportunity to observe the subsequent evolution of the introductory physics course. I have, therefore, little knowledge about the effectiveness of the published lectures with later students. It had always been clear that *The Lectures,* by itself, could not serve as a textbook. Too many of the usual trappings of a textbook are missing: chapter summaries, worked out illustrative examples, exercises for homework, and so forth. These would have to be provided by industrious instructors, and some were provided by Leighton and Rochus Vogt, who took responsibility for the course after 1963. I had at one time contemplated that these might be provided in a supplementary volume, but it never materialized.

In my travels in connection with the Commission on College Physics I would often meet with physics faculties at various universities. I would hear that most instructors did not consider *The Lectures* suitable for use in their classes—although I did hear from some who were using one or another of the books in an "honors" class, or as a supplement to a regular text. (I must say that I often got the impression that some instructors were wary of trying *The Lectures,* because of fear that students would ask questions they would be unable to answer.) Most commonly, I would hear that *The Lectures* were found by graduate students to be an excellent source of review for qualifying exams.

It appeared that *The Lectures* may have been making more of an impact in foreign countries than in the United States. The publisher had arranged for *The Lectures* to be translated into many languages—twelve, as I recall. And when I would travel abroad for conferences on high-energy physics, I would often be asked whether I was the Sands of the red books. And I heard frequently that *The Lectures* were being used for courses in introductory physics.

Another unfortunate consequence of my leaving Caltech was that I could no longer keep up my active association with Feynman and his wife Gweneth. He and I had had a cordial collegiality since the Los Alamos days, and in the mid 1950s I had participated at their wedding. On the rare occasions after 1963 when I would visit Pasadena I would stay with them, or when I visited with my family we would always spend an evening together. On the last such occasion he told us the story of his most recent surgery for the cancer that not long afterward claimed his life.

It is a source of great pleasure for me that, now, some forty years after they were given, *The Feynman Lectures on Physics* are still being printed, bought, read, and, I would venture, appreciated.

Santa Cruz, California
December 2, 2004

Interview with Richard Feynman

From *Interview with Richard Feynman by Charles Weiner in Altadena, CA, 4 March 1966*, courtesy of Niels Bohr Library & Archives, American Institute of Physics, College Park, Maryland, USA.

Feynman: *The Feynman Lectures on Physics*. Do you want to talk about that?

Weiner: I think it's appropriate, because this was a very major activity in this period.

Feynman: Yeah. It's interesting, now that I think about it, that since that was a major activity in that period, I was complaining that I wasn't doing any research—I'm really crazy! People have pointed out to me now that it's really quite silly of me to feel that I wasn't doing anything in these years, because that thing [*The Feynman Lectures on Physics*] is something. But I still don't feel it that way, because when you're young, you dedicate yourself to some ideal—that you're going to discover things in physics—and if you do something else, it's hard for you to rationalize that it should satisfy anybody—it's just that I was teaching a class.

So anyhow, the story of those lectures is the following. There was a discussion by some group, of which I was not a member, that they ought to revamp the physics course, because many of the students who were pretty good, who were taking physics, were complaining that after studying it for a year or two, all they were doing was pith balls and inclined planes. They had heard so much when they were in high school of relativity and strange particles and wonders of the world, and they would see nothing of the wonders of the world until they were graduate students. And this was very difficult, and they were trying to revamp the physics course. So they had worked out some kind of a syllabus for it, and the question was, who was to give it? I don't know how they discussed among themselves, but anyway, Sands came over here and he talked me into giving the course.

However, I threw away the syllabus. You know, I decided to give it my own way, of course. But I got the general idea of what was involved.

They wanted me to teach the freshman lectures. They wanted to revamp the course. It used to not have any main lectures by a main lecturer, but they used to have sections, with graduate students teaching different sections. The only thing they ever came together for then was an optional sort of cultural lecture that was not directly related to the course, once a week, on Fridays, or maybe once every two weeks on Fridays.

Weiner: Some historical thing, perhaps?

Feynman: Well, it would be different things. I would often be invited to talk there, and I would talk about relativity. It was not part of their course. Sometimes people would talk about something that was directly part of their course, but it wasn't organized together.

Now they're going to do a new laboratory. They were going to cook up a new lab, and they were inventing new experiments to go with the lab. They were going to redesign it, so that there would be at least two lectures a week given by a main professor, and then some recitation sections that graduate students would pay attention to, and would I give the lectures, see? They had money from the Ford Foundation for this revamping. There's a lot of money around for changing the world these days.

So I said, "Okay." I accepted the challenge for one year, and I tried to make a course that required giving two lectures a week.

Weiner: Didn't you have to drop all other work, all other teaching?

Feynman: I did, in fact. I can hardly believe it, but my wife tells me that I was working essentially day and night, sixteen hours a day, all the time. I was down here all the time, worrying about these—working on these lectures, because I not only had to prepare the material, I also had to prepare the lecture so it was a good lecture, if you know what I mean.

I had the idea—I got a kind of principle, a number of principles. The first was that I wouldn't teach them anything that I had to teach over again because it was wrong, unless I pointed out that it was wrong. For example, if Newton's laws are only approximate, and they're not good in quantum mechanics and they're not good in relativity, I start out by saying that, so that they know where they are. In other words, there always should be some kind of a map. In fact, I even thought of making some sort of a great map of things with their interconnections, so we could see where we were. I thought that one of the troubles with all the courses in physics was that they just said: You learn all this, you learn all that, and when you come out the other end, you'll understand the connections. But there's no map, "guide to the perplexed," you see. So I want to make a map. But it turns out it's not a feasible design. I mean, I just never made such a map.

The other thing is, I wanted to have in it things that would be enough for a good man to chew on, and then also that the average guy should understand. So I tried to invent.

Let me go over the principles. The first was, I'd never introduce anything that was not exactly right without explaining that it wasn't, and what changed next time. (Another thing was, I looked at books, you see, and I began to realize great weaknesses: for example, like they were teaching in the same book $F = ma$, and a little bit later that the frictional force is the constant of friction times the normal force . . . as if they were of the same caliber and the same significance. They're so different in quality and, you know, nothing is made of it.) So that was what the first principle was.

The second principle was: that which *is* supposed to be understandable, and that which is *not* supposed to be understandable from what you've already said, should be made clear. Because I would find in books that they would give all of a sudden, say, the formula for the frequency of an AC circuit. That was supposed to be more advanced. They can't derive it now, but they wouldn't say, "You aren't going to be able to understand this formula at this level with the reasoning that has just preceded, but it's an added thing." In other words, what's being added in, and what should have come from the other thing? Even if it could have come from the other—but you don't make the argument—you should say it. I always say, "This is a possible deduction, more or less as follows, but we haven't tried to deduce it from that." Or, "This is an independent idea that comes from another place, you see, and you can't deduce it, so don't worry."

A few little principles like that. Then the problem was to make lectures which would be okay for the average student, and yet have stuff for the advanced student. Then I got an idea, when I was planning these lectures. I would have a cube in the front of the lecture hall which had different colored faces, so that when something was only for the fun of it, for the more advanced student to get him interested, but wasn't really an essential part of the course, it would be one color face. You see? When there was something that was so basic that it was absolutely necessary to understand for the whole of physics, and everybody should try their best to understand this thing, another color face, and so on. A color face to indicate the importance, the position, of the different subjects. Because what I was worried about was that all the students would try to learn all this junk, and if they do that, then I haven't got stuff for the advanced student.

You can't do it. It's just impossible, to have stuff for the advanced student without possibly confusing the stupidest student or the less advanced student.

So I had this cube idea. But I gave that up as being gimmicky, and instead I would write, at all lectures, summaries (which are no longer extant) on the blackboard, of the central items which needed to be understood. Anything else that wasn't in the summary was just for the fun of it. But those don't exist anymore.*

Finally, let's see—I thought of some other things while I was talking. I don't know.

So, then I started to give the lectures. And at the very beginning, the first thing I wanted to do was get all the students together. At a number of lectures, people don't understand the logic at the beginning. The real logic of the beginning is, get all these kids from high school to come approximately to the same rough position. For instance, I would talk about everything being made out of atoms—not because I think they don't know that, but because I want those who don't know it to know it. I can't say that, you see, so I tell it in such a way that the ones who already know it are excited by it, because it's a new way of looking at it, while the ones who don't know it can just catch onto it, to the level that I need. And so on. So the first few lectures are to bring everybody together.

Also, these lectures were lectures I had given in other places, the beginning lectures especially, so that I could have time to prepare the later ones, you see. And finally—oh, another principle, a very important principle: I wanted each lecture to be able to stand by itself. I didn't think it was a good idea to have a lecture and say, "Well, the hour is up; we will continue this discussion next time," or "Last time when we left off, we were doing this, that, and the other thing. Now let us continue."

So, instead of that, I wanted to make believe to myself that each lecture was somehow or other an isolated masterpiece, you see, of lecturing, in which you had a beginning, and introduction, and you had a conclusion with some drama. So each of the lectures were like that, with some minor exceptions. There were one or two places where I couldn't do it, where I continued the two lectures together or something like that—but that was another principle. I'm just telling you the guides that made those things.

Finally, my main interest is in physics, and in organizing material. I love to organize the material, and to think about how it goes together,

---

* Feynman's lecture summaries are preserved in blackboard photos kept at the Caltech Archives; they will be published in enhanced electronic editions of *The Feynman Lectures on Physics*. See http://www.basicfeynman.com/enhanced.html.

and to discover a new way of looking at something, and how I can explain it and so on. And I'm not the kind of a teacher who's interested really in the student as an individual. I mean, I'm not worried about: this guy's married and he's trying to get his degree, and all these complications. I tried my best to teach the student more or less as the abstract student, with imaginary properties—mixed, mixed, there were many different kinds of abstract students—but not any particular individuals. The subject is the center of my interest in all cases—the subject, not the student but the subject. So, you want to know how I feel about them [the lectures]. What else can I say about them? They're all published. But I'm trying to explain to you how I feel, myself, about them, and what I thought I was trying to do.

Weiner: Did you get any sense of feedback while you were doing it?

Feynman: No. None whatever, because I had no way to know what was happening. Because I didn't have any recitation sections, and I didn't have questions at the end of the lecture. Any questions were supposed to go into the recitation sections. So there was zero feedback, except that there were some exams in which people made up problems. They gave them problems, and they would try to write the answers, in certain exam weeks, you see. And they were so atrocious—as far as I was concerned—they were so zero that I really, in a certain sense, felt discouraged throughout the entire program. Not discouraged to the point of not keeping right on going the way I was going, but in the feeling throughout that it wasn't working, that it's useless—but never mind, I'll do it anyway. I mean, it's the only way I know how to do it, damn it. But it doesn't work.

Weiner: How about the people who were directly in contact, with the recitations?

Feynman: People who were directly in contact would tell me that I was underestimating them, and that it wasn't as bad as I thought. But I never believed them and still don't.

Weiner: Don't you think that this type of presentation, the effectiveness of it, is difficult to measure in a traditional examination?

Feynman: Of course it is. But let's just assume that you're getting somewhere. But what else do you do? I mean, you asked me what my reaction was. It may be difficult, but I expected them to do better on the simple questions than they were doing. In other words, a person who couldn't do what they apparently couldn't do was certainly not understanding what I was talking about. That's the way I felt about it.

Weiner: How long did you do this? Three years?

Feynman: I did this for a year, and then they started to work on me for the second year. And I said, "I prefer to do the first year over again. This time I want to make up problems that go with the material, and to make some improvements, but mainly to make up problems to go with the material, so that it would really teach it." And to make some improvements of things I didn't care for.

Then they worked on me, and I'm glad they did — in some way, anyway. They said, "Look, nobody's ever going to do this again. We need this second year."

I didn't like to do the second year, because I didn't think I had great ideas about how to present the second year. I felt that I didn't have a good idea on how to do lectures on electrodynamics. But, you see, in these challenges that had existed before about lectures, they had challenged me to explain relativity, challenged me to explain quantum mechanics, challenged me to explain the relation of mathematics to physics, the conservation of energy. I answered every challenge. But there was one challenge which nobody asked, which I had set myself, because I didn't know how to do it. I've never succeeded yet. Now I think I know how to do it. I haven't done it, but I'll do it someday. And that is this: How would you explain Maxwell's equations? How would you explain the laws of electricity and magnetism to a layman, almost a layman, a very intelligent person, in an hour lecture? How do you do it? I've never solved it. Okay, so give me two hours of lecture. But it should be done in an hour lecture, somehow — or two hours.

Anyhow I've now cooked up a much better way of presenting the electrodynamics, a much more original and much more powerful way than is in the book. But at that time I had no new way, and I complained that I had nothing extra to contribute for myself. But they said, "Do it anyway," and they talked me into it, so I did.

When I planned it, I was expected to teach electrodynamics, and then to teach a subject which would really be all the different branches of physics, using the same equation — like you use a diffusion equation for diffusion, for temperature, for lots of things, or the wave equation for sound, for light, and so on. In other words, the second half would have been something like mathematical methods of physics, but with many physics examples, so I'm teaching physics at the same time as the mathematics. I would teach Fourier transform, differential equations, and so on. It wouldn't look like that, though. It wouldn't be organized the usual way. It would be in terms of subjects, the point being that the equations are the same in so many different fields. So the moment you deal with an equation, you ought to show all the fields that it comes from, instead of just talking about the equation. So I was going to do that.

But then I had another possibility. Maybe I could teach quantum mechanics to the sophomores; nobody expects that to be done—that would be a miracle. And I had a crazy upside-down way of presenting quantum mechanics, absolutely inside out, in which everything that was advanced would come first, and everything that was elementary, in the conventional sense, would come last.

And I told these guys about that, and they kept working on me. They said I had to do it, that the mathematical thing that I was talking about, other people may some day do, but that this thing would be so unique, and they knew that I would never go for another year. I must do this unique thing, you see—even if it kills the kids, they can't learn it, and it's no good. I don't know what the situation is, actually, whether it's worthwhile or not. I should try it. So I did. And that's Volume 3 on quantum mechanics. But Volumes 2 and 3 were really one year, just like Volume 1 was.

Weiner: This represents two full years you put in.

Feynman: Right. Well, one is '61–62, and the next is '62–63.

Weiner: And since then, of course, as you mentioned yesterday, you have better feelings about it . . .

Feynman: Somewhat.

Weiner: Because of their use beyond Caltech.

Feynman: Well, I haven't yet, but people have pointed out I ought to. And I may be gradually coming around to understanding that. But what I insisted that I was doing, from the beginning, was teaching this particular group of students, and that's all that I could do. I kept saying, "You cannot live beyond the grave. You teach these students, that's all it's going to be, and there won't be any way to get this to anybody else." I think it's roughly true. If I listen to the lectures that other people give, on the basis of these books, I see all kinds of flaws, errors, weaknesses, and distortion. And it is true that you can't live beyond the grave. But there must be people living who aren't listening to the lectures of some professor, who are sitting just reading the book and thinking for themselves. They must get something out of it. So, if I keep some hope that that's worth something to them, maybe I can feel better about the whole thing. I think that, in regard to the particular students that I was really aiming at, which was my avowed purpose that I'd set—I wasn't caring about the books or anything, I was only caring about the students—I think that the result was nowhere near worth the effort.*

---

* Two decades later, speaking of *The Feynman Lectures on Physics*, Feynman said, "There is all kinds of stuff there, more basic physics points of view, and so apparently they are useful. I must admit now that I cannot deny that they are really a contribution to the physics world."—from J. Mehra, *The Beat of a Different Drum* (1994)

Interview with
Robert Leighton

From *Robert Leighton Oral History Interview with Heidi Aspaturian in Pasadena CA, 8 October 1986,* courtesy of the Caltech Archives, California Institute of Technology, Pasadena, California, USA.

Leighton: The Feynman course was important, and I played a role there in the editing, and translating "Feynmanese" into English. That was an interesting and exciting time.

In the early 1960s, when [Gerry] Neugebauer and I were talking about infrared, and when I was getting interested in Mariner, along came *The Feynman Lectures*. That resulted from a project—in which I played some direct role—to redo the freshman physics course. I had some ideas about how to do that, and some of the other people on the freshman physics committee had some ideas as well. But partway through the discussions, Matt Sands said, "Well, really, we should have Dick Feynman present the lectures and have them tape-recorded." Sands was then a professor of physics at Caltech. He was a very forward type of fellow. He had been on the Los Alamos project as a young person, so he knew Feynman well enough to go and talk to him. But Feynman resisted.

Aspaturian: What was it about Feynman's lectures that made him the obvious choice for this kind of thing?

Leighton: Feynman has a peculiar property, which is that at the time he's explaining something, it appears very clear and transparent—you can see how everything fits, and you go away feeling very good about it, as if, "Well, there's a lot of loose ends there that I want to follow up on; but boy, wasn't that great!" And about two hours later, like what they say about Chinese food, it's all gone and you're hungry again. And you don't remember quite what happened.

I witnessed it myself. In the late fifties, Feynman gave a talk to a lay audience on the basic ideas of Einstein's special theory of relativity in 201 East Bridge—the lecture hall was tremendously crowded, of course. In his characteristic way he reduced the subject to its lowest terms, about the $1 - v^2/c^2$—"all you have to learn about is this square root of $1 - v^2/c^2$."

After the lecture, on the way out, I overheard a young lady saying to her escort, "I didn't understand much of what he said, but it sure was interesting!" Feynman had a way of doing that.

Aspaturian: It sounds like he gave virtual lectures in the sense of virtual particles.

Leighton: [Laughter] Well, that's right. Yes, bringing the thing out into reality only for a limited time, and then watching it sink back into the sea!

Aspaturian: The idea was to get him out of the vacuum permanently.

Leighton: Yes. So Matt Sands went to Feynman, and Feynman balked, but eventually he agreed to do it. And that was where *The Feynman Lectures* came from.

Leighton: In his teaching, Feynman tried to organize undergraduate physics into a two-year sequence, which turned out to be three years, because in the first two years he didn't really get to quantum mechanics—although he did deal with isolated pieces here and there. He started right out with atoms—he didn't hold back on atoms, leaving them to the chemists, and teaching only pulleys and strings to the freshmen! He pushed the freshmen's nose into the fact that what physics *is* is the properties of atoms. In this categorizing way, he tried to make each lecture an independent, self-standing thing. Now, you can only do that to a certain extent, because you've got to base your knowledge on some level of mathematics and on some sophistication in the application of mathematics to physics, and things like that.

Anyway, at first it seemed like a great idea to get Feynman to do this. As a matter of fact, it turned out to be a better deal for mature physicists than for the freshmen. Feynman's course was a little too rich for most of our freshmen: for about 20 percent, it was the ideal thing, absolutely great; for about 60 percent, it was not. Their reaction was more like, "Exactly what do they expect us to learn about in all this?"

I was in charge of the laboratory and the coordination of the course for that first year. I was also in charge of the transcription of the lectures into written form. I explain in the foreword to the book how we expected that the editing was going to be a job for a graduate student—to dot some i's and cross some t's, and change a word here and there that the transcriber might have misunderstood or something.

Aspaturian: How did you happen to get the assignment of overseeing the editing?

Leighton: I'd been chairman of the course modification group. You don't want to hand it to Feynman to run the whole course himself; he's going to give the lectures, and it takes all of his time to do that. There also had to be laboratory experiments to go with them, and the new material was sufficiently different to call for quite different experiments in the freshman lab. Dr. [H. Victor] Neher, who is now retired, really was in charge of the laboratory part. But I was the coordinator.

The lectures were taped; Feynman used one of those cordless lapel microphones, and we hired a young lady to transcribe them. She was just as happy as could be, listening to that material and typing it. She did a fine job. But about six or eight lectures went by and nothing usable came out the other end. The transcript was verbatim, and in this case verbatim is bad—because Feynman never says anything once: he says it at least two-and-a-half, if not three-and-a-half or four times—and he puts it a different way each time. Then he'll go on to the next topic for another couple of minutes, and he'll still be thinking about whether he could have explained his earlier topic better, and then he comes back. The results were loosely organized, modestly disorganized. I wound up, myself, personally, doing the editing for the first volume. It was a full-time job; you couldn't present the material successfully without paying very careful attention to it.

There's one particular passage, which I'm sure I could find if I looked in the Feynman book. I'd like you to see what form it was in when it first came out of Feynman. [Laughter] It had to do with physics before Newton and physics after Newton. Feynman's point was that, before, the world was just a tremendous confusion of darkness and superstition— and afterwards, it was all light and structured and understandable. It was absolutely true, but he was trying to say this in a way that never did quite gel. He had a sentence in there that never had a verb in it! [Laughter]

Aspaturian: How well did you know Feynman when you started?

Leighton: Oh, about as well as I know him today. I guess he and I share a certain aspect of social ineptness: I can't remember people's names unless I study them very carefully, and for quite a long time. If I want to catalogue somebody's name in my head so I can get it again, I have to do it right then. But the trouble is, I'm introduced to somebody just in the middle of a conversation, and the conversation goes on—and who he or she is just drops out of my mind. It's one of those handicaps; Feynman has it, too. He roomed for at least a term at MIT, I believe, with somebody who was later at Caltech, and he couldn't remember his name! [Laughter]

Aspaturian: What was it like working with him on *The Lectures*?

Leighton: What initially came out in the transcript was absolutely raw "Feynmanese" that had to be rough-edited right on the original sheets. After I got his material from each lecture into a form that I thought was ready for typing on the master sheet, it was sent back to the young lady and rendered into a form where it could be shown to Feynman. He would look at the thing now and then, but usually had no comment—that is to say, he was sufficiently satisfied with it.

Another thing is that the lecture came at eleven o'clock, followed by lunch. We would walk to lunch together, and when he was dissatisfied about the way something or other was worked out, there would be questions or comments about, "What could we do to do it better?" There would be ideas and we'd talk. There'd be other people at the lecture—professors and TAs—and there would be sort of a floating lunchtime, which was partly devoted to just talking about that lecture. It was not structured consciously that way, but it was an opportunity to get some ideas.

Aspaturian: Was this originally designed mostly for the benefit of Caltech students?

Leighton: Oh, yes.

Aspaturian: But then it sort of spread out, didn't it?

Leighton: Well, no physics instructor who was teaching freshman physics could resist having a copy of *The Feynman Lectures,* whether or not he was using it in his class. This project was financed by a Ford grant, and I don't know what the royalty figure has come up to. It was an arrangement where the Institute agreed to put any royalties that the texts might accrue into support for similar kinds of activities at Caltech. None of the royalties went to any of the people involved with the lectures themselves. These were academic assignments, so the project was not treated as a copyrighted manuscript. It was just as well. At the time, Feynman said, "We will know whether it sells very well by seeing how big our salaries are in the next four or five years." [Laughter] And he was right. Our salaries went way up—his, for obvious reasons, and a lot of the rest of us because of being nearby, I guess.

Aspaturian: Your son Ralph got involved in doing something similar.* How did that happen? Has this become a family privilege?

---

* Ralph Leighton was Feynman's amanuensis for two collections of reminiscences— *Surely You're Joking, Mr. Feynman!* (Norton, 1985), and *What Do You Care What Other People Think?* (Norton, 1988), which were combined into a single volume, *Classic Feynman*, in 2005.

Leighton: I can't quite remember the order in which things happened, but my wife and I would have dinner parties, and Feynman must have come to one or more of them. My son Ralph was at that time in high school and interested in drumming, and he was friendly with a very musical family in which there were lots of kids and parents who played various instruments—that would bring another group of visitors to our house. On one of these occasions, Feynman heard Ralph and his friends drumming at the other end of the house and, of course, he went in—he was more comfortable with kids anyway. He introduced himself and they invited him to drum. And that led to rather regular drumming sessions by Feynman, Ralph, and a couple of other drop-in friends.

I myself was curious about Feynman's drumming ability, so I asked Ralph one time, "Well, how good a drummer is Feynman?" He said, "Well, he picks up the rhythms all right, and he's very fast, but sometimes he has a hard time getting started—but for an old guy, he's pretty good." [Laughter] I informed Ralph that he had just spoken of the capabilities of possibly the one person in the world who knew more about how everything in the universe worked than anyone else on Earth at that moment. [Laughter]

Anyway, Ralph's other musical friends gradually went off to college here and there, but Feynman and Ralph continued drumming together. If you were around Feynman long enough, you'd hear these amazing stories in some random order. Undoubtedly, they gain with the telling, but they're all quite real. There's an infinite cauldron, out of which he'd dig one of them up on occasion. That is to say, something in the conversation would recall such-and-such. If you happened to have been near him during some similar conversation, you might have heard the same story—Feynman fixing radios as a kid, or interacting with generals at Los Alamos, for example. And Feynman, he can go on forever: one thing reminds him of another—it's amazing. The man is absolutely incredible.

Aspaturian: So, there's an inexhaustible store of lore there.

Leighton: Or, as some people would say, inexcusable! [Laughter]

During their drumming sessions, Ralph made tapes of these stories. Then he transcribed them—first on a typewriter, and then on my computer. Feynman was in favor of this; it was not surreptitious at all. It was simply Ralph saying, "These stories are so great, but they're like gems slipping through my fingers—can I tape them?"

Then at some stage, I said to Ralph, "How about running the transcripts by me? I'd just like to refresh my memory." So I read most of them. Now and then I would see some word that was misunderstood.

Aspaturian: You were familiar with most of them?

Leighton: Oh, yes. Only about 20 percent were new to me. I think Ralph and I, without ever discussing it, on our very different projects, realized the same thing about Dick: namely, you should do a minimum of editing on what he says. You should leave it as close to the original as possible, including the mannerisms—although not the repetition. In the physics lectures, I found it absolutely essential to crunch the repetitive material down into what might be a good way to put it, and then let it go at that. Ralph has a lot of talent along those lines. However, that particular job was the first time he had ever tried to write something for publication, so he got some valuable lessons on editing from [Engineering and Science editor] Ed Hutchings.

Aspaturian: Is there a sequel planned?

Leighton: Well, there are still more stories. And then there's also *QED* [*QED: The Strange Theory of Light and Matter* by Richard Feynman (Princeton, 1985)] which has come out and has gotten pretty good reviews. And I guess Ralph is still running the tape recorder.

Aspaturian: There were a few things in that book [*Surely You're Joking, Mr. Feynman!*] that I found did not reflect terribly favorably on Feynman. Was there any discussion about getting rid of some of those?

Leighton: No. That's the man.

Interview with
Rochus Vogt

The material for this section was recorded by Ralph Leighton on May 15, 2009, at the California Institute of Technology. Leighton and Michael Gottlieb interviewed Rochus E. (Robbie) Vogt about Caltech in the early 1960s, and what it was like to teach Feynman's physics. (Exclamation points usually indicate that Vogt was laughing about what he was saying at the time.)

Leighton: I'd like to ask you about your role in *The Feynman Lectures on Physics*. Take us back to those days, if you would.

Vogt: I came to Caltech in 1962, and the freshman course was given in '61—so I came in the first year that Feynman's freshman course had to be translated into something that common people could do—and it was quite a challenge! When Caltech hired me, I told Carl Anderson, the physics department chair, "I've got to finish up some important work in Chicago, and I cannot get away until mid-October." He said, "No problem; somebody will take your class until mid-October, but as soon as you show up, you teach!" It was very different from the way it is today. I remember my wife, Micheline, and I arrived in Pasadena on a Saturday afternoon, and Monday morning, I was in a classroom—and I didn't know what I was doing!

It was the second year of the course. Feynman gave the sophomore lectures, while your father [Robert Leighton] took over the freshman lectures. Leighton gave very good lectures, and it was very enjoyable to work on that team—and also it was great to see whether we mortals could teach Feynman, which many people doubted was possible! Under Bob Leighton, I was a Teaching Assistant (TA) teaching two recitation sections—a common one, and the honors section. The honors section was kind of neat; the common one wasn't as neat, because it had biologists in it who didn't want to learn physics! Nevertheless, it worked out. It was more of a challenge than the honors section—the honors section was so much easier to teach: they did it all on their own; they didn't need me.

Leighton: It's funny how you can think you're a good teacher—when you have good students!

Vogt: That's right. At that time there was a TQFR—a Teaching Quality Feedback Report—on all the faculty, which went on all the time, and I read my own. It said, "He's doing a very good job, but of course anybody could do that with a good textbook like *Feynman!*" So they thought it was a good textbook at the time. In later years, people at Caltech said *The Feynman Lectures* are really not suitable as a textbook—but it's amazing how many people read it in parallel with something they have been assigned—and that means *The Lectures* haven't been lost. But at Caltech, it should still be the text, period!

It wasn't easy, because none of us had the charm or allure of Feynman—nobody can imitate that. But in my second year, when I gave the freshman lectures (following Bob Leighton), I always gave this assignment: read the following chapter of *Feynman*, and then I will teach you *what to do with it*. That worked, because I did not parrot Feynman. In fact, I said to them, "There's no sense in me trying to parrot the Bible—it stands on its own feet—but I can tell you how to work with it." I gave them examples, applications, amplifications, and sometimes interpretations—because Feynman sometimes was on a pretty high level—and that seemed to work.

You might be amused how I took over the *Feynman Lectures* in my second year at Caltech. One day, in early October, Bob Leighton and I ran into each other, and he said, out of the blue, "Robbie, I want you to take over the class."

"What's going on, Bob?" I said with concern.

He says, "I need a sabbatical, and I've decided I'm going to go to Kitt Peak in Arizona, and I've decided you'll take over the Feynman course." So word got out that Bob Leighton was planning to pass *The Feynman Lectures* on to me.

Matt Sands, when he heard about it, went through the roof! I remember talking about it with Bob Leighton in his office, and outside Matt Sands was yelling in full voice at no one in particular, as far as I could tell, "Bob Leighton has gone insane! He's crazy! He's letting this inexperienced green assistant professor take over the Feynman course! This is an outrage! I protest!" He was really agitated, because he cared deeply. He trusted Bob Leighton, but he had never heard of me.

Anyway, I gave my first lecture for the Feynman course on the 21st of October, 1963. Several things had happened: I was going to a conference in India during quarter break in December, so had just gotten some yellow fever shots and some typhoid shots—and when I got the typhoid shots, I developed a high fever—and so on the 20th of October I had a high fever. On top of that, my wife, Micheline, brought our first daughter, Michele, into the world on that day, so I spent the night of October 20 in the hospital, waiting for things to happen! So I had only a couple of hours of sleep, I had a high fever, and I gave my first Feynman lecture—it was quite a start.

Incidentally, your mother, Alice, did something absolutely marvelous: she called us up and said, "I feel bad that Bob got you stuck with the Feynman course, and I know you kids are just getting going in life, so I have decided to subscribe you to a diaper service—this will give you some help," which it was.

Anyway, as I said, I was very comfortable teaching Feynman physics, because these were very bright students: if you gave them a break, they would do something good with it. I think they were actually more able to do things under me than they were under Feynman, because in addition to Feynman, they got somebody who gave them *applications* of Feynman.

As you may know, more than half of the TAs were professors when Feynman was lecturing. But even when I was the lecturer, there were several professors with recitation sections—one of my TAs was Tommy Lauritsen. Tommy was very helpful. He came to each lecture and told me whether it was good, or whether it could be improved. Being a TA was considered necessary preparation to give *The Feynman Lectures;* after I gave *The Lectures* for two years, Tommy took over from me—he was the next *Feynman* lecturer.

When I was teaching the recitations under Bob Leighton, I got very familiar with the Feynman course. Otherwise, if I had walked in cold without that background, I couldn't have done a good job. As a TA I had learned what the students needed—what worked with them, what didn't work; even when I was the lecturer, I always taught a recitation in parallel with my lectures, because I wanted to find out how the students were doing, and what I could do better. When you are in a small class with ten to twenty students, you get very good feedback, whereas as a lecturer you get little feedback, because they're so busy taking notes and listening. Sometimes you stay after class a little bit, but it's not the same thing. But when you give them homework, and discuss it with them, you find out whether the students actually can do the physics.

I had a philosophy about homework that is in contrast to what they're doing now—i.e., they print the solutions and hand them out to the students the day the homework is due, or use last year's printout, because they often use the same problems again. I am totally opposed to that. It's psychological: when you get stuck, and you absolutely don't know what to do next, you naturally would like to look at the solution to get you over the hump. But pretty soon you begin to look at the solutions earlier and earlier. And so I made my philosophy very clear to the students. I said, "I expect you to try first to do your homework alone. But if you spend twenty minutes on a problem which I assign to you, and you still don't know how to do it, then go and talk to others. Don't feel bad about it. Sometimes you just don't get it; you may have missed something critical. As soon as somebody gives you the word, you know how to do it. However, once you understand the problem, go back to your room and write up the solution *on your own*—don't copy somebody else's solution."

There was a third phase: I said, "If as a group you can't do it after half an hour, call me." I had forgotten when students do their homework—thus I got phone calls at two or three o'clock in the morning: "We're stuck! We have spent the last *hour* and we're getting nowhere!"

Gottlieb: I would have given them another problem: "What is the latest hour that you think it is decent to call a professor?" [Laughter]

Vogt: Actually, I was grateful that they tried to do it. And when you're young, it's not a big deal to wake up at three in the morning, spend fifteen minutes talking to some students, and go back to sleep—especially when you've got a baby crying in the other room anyway! At least I knew what to do about the students' problems; the baby's crying, I had no *idea!*

Going back to your first question, Ralph, about my role in the Feynman course: I saw myself as an acolyte who was an interpreter, a go-between between Feynman and the students. Another role I had was coming up with exercises, along with Bob Leighton. He was very influential: I mean, he made *me* do it! He often said, when we created the A, B, and C problems, "We need a couple of more As, or a couple of more Bs." Usually we had plenty of Cs, the hardest ones! He always knew what was missing. Sometimes he came up with a problem, but very often, he said, "Robbie, go think up a couple more problems—I know you can do it." That was his style: he felt everybody was competent enough to do things; they just needed motivation to do it. He didn't think he was imposing upon me; he just thought he'd help me to do the right thing!

One time years later I "cheated" and used someone else's problem. There's an important paper by one of my heroes, Val Telegdi, about the calculation of the g-factor of an electron. It was in *Nuovo Cimento* (the Italian physics publication), sixty-five pages I seem to remember, mostly mathematics above my head. I looked through the goddamn paper, and said to myself, "That's so much hard work to go through it!" However, I remembered Feynman's sophomore year lectures on quantum mechanics, and I knew that you could solve that same problem with *Feynman Lectures* physics. So I gave to my junior-year students this homework problem: "Calculate the g-factor of the electron."

More than half the class was able to do it. Now, that was a little bit shifty, because you cannot use the style of quantum mechanics Feynman taught in the sophomore year for everything, but it has great applicability to certain physics problems like this one. You have no idea how proud the students were: in one and a half pages, they could do a piece of physics that took Telegdi sixty-five pages and a lot of math! And so they thought Feynman quantum mechanics was very elegant, which it is.

**

Another thing I remember, coming back to my early years, when we taught the Feynman course: Every week, on Wednesdays, some six to ten physicists had lunch together (we brown-bagged it, or went to Mijares Mexican restaurant in Pasadena), which included Bob Leighton, Gerry Neugebauer, Tommy Lauritsen, and others. When we got together at those lunches, we talked about teaching: what worked, what didn't work, what we could do better. There was so much mutual support that you could become a better teacher, because you had all that help—also on Friday afternoons at the Lauritsens, where a lot of us would unwind with martinis at the end of the week. We mostly talked about students and teaching. We talked about research at other times, since we were each doing different research areas, and we had different opinions on how fascinating things were that somebody else did—each of us thought our own research was the most fascinating, of course—but when it came to teaching, we were interested in what everybody else did, because we could learn from them. Nobody forced us to do that; it just happened spontaneously in the climate of Caltech in the early 1960s.

That's how the Feynman course arose, I understand—at the Lauritsens over drinks. They were talking about how to do things better, and Matt Sands came up with the idea that they should rope in Feynman.

It was at such get-togethers that I saw how a university can be a very rewarding and warm place—because of the students: they form a link between the faculty. We got together because of the students, not because of our research. Of course, we also got together individually—Tommy often came to my lab and said, "Tell me what you're doing," and had good suggestions, but this was a one-on-one usually. This student business was a collegial activity. When I gave my lectures, there were usually three or four professors sitting in the back row, in 201 East Bridge, the big lecture hall—*not* because they didn't trust me, or spied on me, but because they were curious how I was doing it, and what could be learned from it. Even Carl Anderson, the division chair, attended every other one of my lectures, and I got feedback from everybody. That was the Feynman spirit: you see, when Feynman taught the course, it was *full* of professors in the back row. They were so fascinated. And so they got in the habit even to come to the lecture of a common person—a boring person like me—because it had become a pattern. This is important. That's what I regret: I don't see that spirit today.

One last thing: in those days I was responsible for my lectures. I assigned all the homework, I made up all the quizzes, I made up all the final exams—personally—nobody else did it *for* me. I wouldn't have asked somebody else, because I thought I knew better what to ask! In addition to that, I taught an honors section; in addition to that, I ran the freshman lab—that was a normal teaching assignment in those days. Today, I think it is one quarter of that. Most professors today teach one class for two quarters per year. Now, I am a fair person: I recognize that today, what we did then would no longer be possible, because today, professors have to spend so much time raising money for their research and defending their research—but that's another story.

1 *Prerequisites*

REVIEW LECTURE A

1-1 Introduction to the review lectures

These three optional lectures are going to be dull: they go over the same material that we went over before, adding absolutely nothing. So I'm very surprised to see so many people here. Frankly, I had rather hoped there would be fewer of you, and that these lectures wouldn't be necessary.

The purpose of relaxing at this time is to give you time to think about things, to piddle around with the things that you heard about. That's by all odds the most effective way of learning the physics: it's not a good idea to come in and listen to some review; it's better to make up the review for yourself. So I'd advise you—if you're not too far lost, completely befuddled and confused—that you forget about these lectures and piddle around by yourself, and try to find out what's interesting without grinding down some particular track. You'll learn infinitely better and easier and more completely by picking a problem for yourself that you find interesting to fiddle around with—some kind of a thing that you heard that you don't understand, or you want to analyze further, or want to do some kind of a trick with—that's the best way to learn something.

The lectures that we have been giving so far are a new course, and have been designed to answer a problem we presumed existed: nobody knows how to teach physics, or to educate people—that's a fact, and if you don't like the way it's being done, that's perfectly natural. It's impossible to teach satisfactorily: for hundreds of years, even more, people have been trying to figure out how to teach, and nobody has ever figured it out. So if this new course is not satisfactory, that's not unique.

At Caltech we are always changing the courses in the hope of improving them, and this year we changed the physics course again. One of the complaints in the past was that the students who are nearer the top find the whole subject of mechanics dull: they would find themselves grinding along, doing problems, studying reviews, and doing examinations, and there was no time to think about anything; there was no excitement in it; there was no description of its relation to modern physics, or anything like that. And so this set of lectures was designed to be better that way, to a certain extent, to help out those fellows, and to make the subject more interesting, if possible, by connecting it to the rest of the universe.

On the other hand, this approach has the disadvantage that it confuses many people, because they don't know what it is they're supposed to learn—or, rather, that there's so much stuff that they can't learn all of it, and they haven't got enough intelligence to figure out what is interesting to them, and to pay attention only to that.

Therefore, I'm addressing myself to those people who have found the lectures very confusing, very annoying, and irritating, in the sense that they don't know what to study, and they're kind of lost. The other people, who don't feel as lost, shouldn't be here, so I now give you the opportunity to go out . . . [1]

I see nobody has the nerve. Or I guess I'm a great failure, then, if I got *everybody* lost! (Maybe you're just here for entertainment.)

1-2 Caltech from the bottom

Now, I am therefore imagining that one of you has come into my office and said, "Feynman, I listened to all the lectures, and I took that midterm exam, and I'm trying to do the problems, and I can't do anything, and I think I'm in the bottom of the class, and I don't know what to do."

What would I say to you?

The first thing I would point out is this: to come to Caltech is an advantage in certain ways, and in other ways a disadvantage. Some of the ways that it's an advantage you probably once knew, but now forget, and they have to do with the fact that the school has an excellent reputation, and the reputation is well deserved. There are pretty good courses. (I don't know about *this* particular physics course; of course I have my own opinion about it.) The people who have come out the other end of Caltech, when they go into industry, or go to do work in research, and so forth, always say that they got a very good education here, and when they compare themselves with people who have gone to other schools (although many other schools are also very good) they never find themselves behind and missing something; they always feel they went to the best school of them all. So that's an advantage.

But there is also a certain disadvantage: because Caltech has such a good reputation, almost everybody who's the first or second in his high school

---

[1] No one went out.

class applies here. There are lots of high schools, and all the very best men[2] apply. Now, we have tried to figure out a system of selection, with all kinds of tests, so that we get the best of the best. And so you guys have been very carefully picked out from all these schools to come here. But we're still working on it, because we've found a very serious problem: no matter how carefully we select the men, no matter how patiently we make the analysis, when they get here something happens: *it always turns out that approximately half of them are below average!*

Of course you laugh at this because it's self-evident to the rational mind, but not to the emotional mind—the emotional mind can't laugh at this. When you've lived all the time as number one or number two (or even possibly number *three*) in high school science, and when you know that everybody who's below average in the science courses where you came from is a complete idiot, and now you suddenly discover that *you* are below average—and half of you guys *are*—it's a terrible blow, because you imagine that it means you're as dumb as those guys used to be in high school, relatively. That's the great disadvantage of Caltech: that this psychological blow is so difficult to take. Of course, I'm not a psychologist; I'm imagining all this. I don't know how it would really *be*, of course!

The question is what to do if you find you're below average. There are two possibilities. In the first place, you could find that it's so difficult and annoying that you have to get out—that's an emotional problem. You can apply your rational mind to that and point out to yourself what I just pointed out to you: that half of the guys in this place are going to be below average, even though they're all tops, so it doesn't mean anything. You see, if you can stick out that nonsense, that funny feeling, for four years, then you'll go out into the world again, and you'll discover that the world is just like it used to be—that when, for example, you get a job somewhere, you'll find you're Number One Man again, and you'll get the great pleasure of being the expert they all come running to in this particular plant whenever they can't figure out how to convert inches to centimeters! It's true: the men who go out into industry, or go to a small school that doesn't have an excellent reputation in physics, even if they've been in the bottom third, the bottom fifth, the bottom *tenth* of the class—if they don't try to drive themselves (and I'll explain that in a minute), then they'll find themselves very much in demand, that what they learned here is very useful, and they're back where they were before: happy, Number One.

---

[2]Only men were admitted to Caltech in 1961.

On the other hand you can make a mistake: some people may drive themselves to a point where they insist they have to become Number One, and in spite of everything they want to go to graduate school and they want to become the best Ph.D. in the best school, even though they're starting out at the bottom of the class here. Well, they are likely to be disappointed and to make themselves miserable for the rest of their lives being always at the bottom of a very first-rate group, because they picked that group. That's a problem, and that's up to you—it depends on your personality. (Remember, I'm talking to the guy who came into my office because he's in the lowest tenth; I'm not talking to the other fellows who are happy because they happen to be in the upper tenth—that's a minority anyway!)

So, if you can take this psychological blow—if you can say to yourself, "I'm in the lower third of the class, but a third of the guys are in the lower third of the class, because it's got to be that way! I was the top guy in high school, and I'm still a smart son-of-a-gun. We need scientists in the country, and I'm gonna be a scientist, and when I get out of this school I'll be all right, damn it! And I'll be a *good* scientist!"—then it'll be true: you *will* be a good scientist. The only thing is whether you can take the funny feelings during these four years, in spite of the rational arguments. If you find you can't take the funny feelings, I suppose the best thing to do is to try to go somewhere else. It's not a point of failure; it's simply an emotional thing.

Even if you're one of the last couple of guys in the class, it doesn't mean you're not any good. You just have to compare yourself to a reasonable group, instead of to this insane collection that we've got here at Caltech. Therefore, I am making this review purposely for the people who are lost, so that they have still a chance to stay here a little longer to find out whether or not they can take it, okay?

I make now one more point: that this is not a preparation for an examination, or anything like that. I don't know anything about the examinations—I mean, I have nothing to do with making them up, and I don't know what's going to be on them, so there's no guarantee whatsoever that what's on the examination is only going to deal with the stuff reviewed in these lectures, or any nonsense of that kind.

1-3 Mathematics for physics

So, this guy comes into my office and asks me to try to make everything straight that I taught him, and this is the best I can do. The problem is to try to explain the stuff that was being taught. So I start, now, with the review.

I would tell this guy, "The first thing you must learn is the mathematics. And that involves, first, calculus. And in calculus, differentiation."

Now, mathematics is a beautiful subject, and has its ins and outs, too, but we're trying to figure out what the minimum amount we have to learn for *physics purposes* are. So the attitude that's taken here is a "disrespectful" one towards the mathematics, for sheer efficiency only; I'm not trying to undo mathematics.

What we have to do is to learn to differentiate like we know how much is 3 and 5, or how much is 5 times 7, because that kind of work is involved so often that it's good not to be confounded by it. When you write something down, you should be able to immediately differentiate it without even thinking about it, and without making any mistakes. You'll find you need to do this operation all the time—not only in physics, but in all the sciences. Therefore differentiation is like the arithmetic you had to learn before you could learn algebra.

Incidentally, the same goes for algebra: there's a lot of algebra. We are assuming that you can do algebra in your sleep, upside down, without making a mistake. We know it isn't true, so you should also practice algebra: write yourself a lot of expressions, practice them, and don't make any errors.

Errors in algebra, differentiation, and integration are only nonsense; they're things that just annoy the physics, and annoy your mind while you're trying to analyze something. You should be able to do calculations as quickly as possible, and with a minimum of errors. That requires nothing but rote practice—that's the only way to do it. It's like making yourself a multiplication table, like you did in elementary school: they'd put a bunch of numbers on the board, and you'd go: "This times that, this times that," and so on—Bing! Bing! Bing!

1-4 Differentiation

In the same way you must learn differentiation. Make a card, and on the card write a number of expressions of the following general type: for example,

$$1 + 6t$$
$$4t^2 + 2t^3$$
$$(1 + 2t)^3 \qquad (1.1)$$
$$\sqrt{1 + 5t}$$
$$(t + 7t^2)^{1/3}$$

and so on. Write, say, a dozen of these expressions. Then, every once in a while, just take the card out of your pocket, put your finger on an expression, and read out the derivative.

In other words, you should be able to see right away:

$$\frac{d}{dt}(1 + 6t) = 6 \text{ Bing!}$$

$$\frac{d}{dt}(4t^2 + 2t^3) = 8t + 6t^2 \text{ Bing!} \tag{1.2}$$

$$\frac{d}{dt}(1 + 2t)^3 = 6(1 + 2t)^2 \text{ Bing!}$$

See? So the first thing to do is to memorize how to do derivatives—cold. That's a necessary practice.

Now, for differentiating more complicated expressions, the derivative of a sum is easy: it's simply the sum of the derivatives of each separate summand. It isn't necessary at this stage in our physics course to know how to differentiate expressions any more complicated than those above, or sums of them, so that in the spirit of this review, I shouldn't tell you any more. But there is a formula for differentiating complicated expressions, which is usually not given in calculus class in the form that I'm going to give it to you, and it turns out to be very useful. You won't learn it later, because nobody will ever tell it to you, but it's a good thing to know how to do.

Suppose I want to differentiate the following:

$$\frac{6(1 + 2t^2)(t^3 - t)^2}{\sqrt{t + 5t^2}(4t)^{3/2}} + \frac{\sqrt{1 + 2t}}{t + \sqrt{1 + t^2}}. \tag{1.3}$$

Now, the question is how to do it with *dispatch*. Here's how you do it with dispatch. (These are just rules; it's the level to which I've reduced the mathematics, because we're working with the guys who can barely hold on.) Watch!

You write the expression down again, and after each summand you put a bracket:

$$\frac{6(1 + 2t^2)(t^3 - t)^2}{\sqrt{t + 5t^2}(4t)^{3/2}} \cdot \left[\vphantom{\frac{6(1+2t^2)}{\sqrt{t}}} \right.$$

$$+ \frac{\sqrt{1 + 2t}}{t + \sqrt{1 + t^2}} \cdot \left[\vphantom{\frac{\sqrt{1+2t}}{t}} \right. \tag{1.4}$$

Next, you're going to write something inside the brackets, such that when you're all finished, you'll have the derivative of the original expression. (That's why you write the expression down again, in case you don't want to lose it.)

Now, you look at each term and you draw a bar—a divider—and you put the term in the denominator: The first term is $1 + 2t^2$; that goes in the denominator. The power of the term goes in front (it's the first power, 1), and the derivative of the term (by our practice game), $4t$, goes in the numerator. That's one term:

$$\frac{6(1 + 2t^2)(t^3 - t)^2}{\sqrt{t + 5t^2}(4t)^{3/2}} \cdot \left[1\frac{4t}{1 + 2t^2} \right.$$

$$+ \frac{\sqrt{1 + 2t}}{t + \sqrt{1 + t^2}} \cdot \left[\vphantom{\frac{}{}} \right. \tag{1.5}$$

(What about the 6? Forget it! Any number in front doesn't make any difference: if you wanted to, you could start out, "6 goes in the denominator; its power, 1, goes in front; and its derivative, 0, goes in the numerator.")

Next term: $t^3 - t$ goes in the denominator; the power, $+2$, goes in front; the derivative, $3t^2 - 1$, goes in the numerator. The next term, $t + 5t^2$, goes in the denominator; the power, $-1/2$ (the inverse square root is a *negative* half power), goes in front; the derivative, $1 + 10t$, goes in the numerator. The next term, $4t$, goes in the denominator; its power, $-3/2$, goes in front; its derivative, 4, goes in the numerator. Close the bracket. That's one summand:

$$\frac{6(1 + 2t^2)(t^3 - t)^2}{\sqrt{t + 5t^2}(4t)^{3/2}} \cdot \left[1\frac{4t}{1 + 2t^2} + 2\frac{3t^2-1}{t^3 - t} - \frac{1}{2}\frac{1 + 10t}{t + 5t^2} - \frac{3}{2}\frac{4}{4t} \right]$$

$$+ \frac{\sqrt{1 + 2t}}{t + \sqrt{1 + t^2}} \cdot \left[\vphantom{\frac{}{}} \right. \tag{1.6}$$

Next summand, first term: the power is $+1/2$. The object whose power we're taking is $1 + 2t$; the derivative is 2. The power of the next term, $t + \sqrt{1 + t^2}$, is -1. (You see, it's a reciprocal.) The term goes in the denominator, and its derivative (this is the only hard one, relatively) has two pieces, because it's a sum: $1 + \frac{1}{2}\frac{2t}{\sqrt{1 + t^2}}$. Close the bracket:

$$\frac{6(1 + 2t^2)(t^3 - t)^2}{\sqrt{t + 5t^2}(4t)^{3/2}} \cdot \left[1\frac{4t}{1 + 2t^2} + 2\frac{3t^2-1}{t^3 - t} - \frac{1}{2}\frac{1 + 10t}{t + 5t^2} - \frac{3}{2}\frac{4}{4t} \right]$$

$$+ \frac{\sqrt{1 + 2t}}{t + \sqrt{1 + t^2}} \cdot \left[\frac{1}{2}\frac{2}{(1 + 2t)} - 1\frac{1 + \frac{1}{2}\frac{2t}{\sqrt{1 + t^2}}}{t + \sqrt{1 + t^2}} \right]. \tag{1.7}$$

That's the derivative of the original expression. So, you see, that by memorizing this technique, you can differentiate *anything*—except sines, cosines, logs, and so on, but you can learn the rules for those easily; they're very simple. And then you can use this technique even when the terms include tangents and everything else.

I noticed when I wrote it down you were worried that it was such a complicated expression, but I think you can appreciate now that this is a really powerful method of differentiation because it gives the answer—*boom*—without any delay, no matter how complicated.

The idea here is that the derivative of a function $f = k \cdot u^a \cdot v^b \cdot w^c \ldots$ with respect to t is

$$\frac{df}{dt} = f \cdot \left[a\frac{du/dt}{u} + b\frac{dv/dt}{v} + c\frac{dw/dt}{w} + \cdots \right] \tag{1.8}$$

(where k and $a, b, c \ldots$ are constants).

However, in this physics course, I doubt any of the problems will be that complicated, so we probably won't have any opportunity to use this. Anyway, that's the way *I* differentiate, and I'm pretty good at it now, so there we are.

1-5 Integration

Now, the opposite process is integration. You should equally well learn to integrate as rapidly as possible. Integration is not as easy as differentiation, but you should be able to integrate simple expressions in your head. It isn't necessary to be able to integrate every expression; for example, $(t + 7t^2)^{1/3}$ is not possible to integrate in an easy fashion, but the others below are. So, when you choose expressions to practice integration, be careful that they can be done easily:

$$\int (1 + 6t) \, dt = t + 3t^2$$

$$\int (4t^2 + 2t^3) \, dt = \frac{4t^3}{3} + \frac{t^4}{2}$$

$$\int (1 + 2t)^3 \, dt = \frac{(1 + 2t)^4}{8}$$

$$\int \sqrt{1 + 5t} \, dt = \frac{2(1 + 5t)^{3/2}}{15} \qquad (1.9)$$

$$\int (t + 7t^2)^{1/3} \, dt = ???.$$

I have nothing more to tell you about calculus. The rest is up to you: you have to practice differentiation and integration—and, of course, the algebra required to reduce horrors like Eq. (1.7). Practicing algebra and calculus in this dull way—that's the first thing.

1-6 Vectors

The other branch of the mathematics that we're involved in as a pure mathematical subject is vectors. You first have to know what vectors are, and if you haven't got a feel for it, I don't know what to do: we'd have to talk back and forth a while for me to appreciate your difficulty—otherwise I couldn't explain. A vector is like a *push* that has a certain direction, or a *speed* that has a certain direction, or a *movement* that has a certain direction—and it's represented on a piece of paper by an arrow in the direction of the thing. For instance, we represent a force on something by an arrow that is pointing in the direction of the force, and the length of the arrow is a measure of the magnitude of the force in some arbitrary scale—a scale, however, which must be maintained for all the forces in the problem. If you make another force *twice* as strong, you represent that by an arrow *twice* as long. (See Fig. 1-1.)

Now, there are operations that can be done with these vectors. That is, if there are two forces acting at the same time on an object—say, two people are pushing on a thing—then the two forces can be represented by two

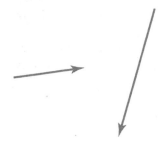

FIGURE 1-1 Two vectors, represented by arrows.

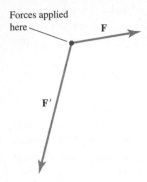

Forces applied here

F

F′

FIGURE 1-2 Representation of two forces applied at the same point.

arrows **F** and **F′**. When we draw a diagram of something like this, it is often convenient to place the tails of the arrows where the forces are applied, even though in general there's no meaning to the location of vectors. (See Fig. 1-2.)

If we want to know the net resultant force, or total force, that corresponds to adding the vectors, and we can draw this by moving the tail of one onto the head of the other. (They're still the same vectors after you move them because they have the same direction and the same length.) Then **F** + **F′** is the vector drawn from the tail of **F** to the head of **F′** (or from the tail of **F′** to the head of **F**), as shown in Figure 1-3. This way of adding vectors is sometimes called the "parallelogram method."

On the other hand, suppose there are two forces acting on an object, but we only know one of them is **F′**; the other one, which we don't know, we'll call **X**. Then, if the total force on the object is known to be **F**, we have

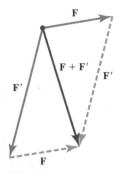

F

F + F′

F′

F′

F

FIGURE 1-3 Vector addition by the "parallelogram method."

$\mathbf{F'} + \mathbf{X} = \mathbf{F}$. And so, $\mathbf{X} = \mathbf{F} - \mathbf{F'}$. Thus to find $\mathbf{X}$ you have to take the difference of two vectors, and you can do that in either of two ways: you can take $-\mathbf{F'}$, which is a vector in the opposite direction as $\mathbf{F'}$, and add it to $\mathbf{F}$. (See Fig. 1-4.)

Otherwise, $\mathbf{F} - \mathbf{F'}$ is simply the vector drawn from the head of $\mathbf{F'}$ to the head of $\mathbf{F}$.

Now, the disadvantage of the second method is that you may have a tendency to draw the arrow as shown in Figure 1-5; although the direction and length of the difference is right, the application of the force is *not* located at the tail of the arrow—so watch out. In case you're nervous about it, or there's any confusion, use the first method. (See Fig. 1-6.)

We can also project vectors in certain directions. For example, if we would like to know what the force is in the '*x*' direction (called the *component* of the force in that direction) it's easy: we just project $\mathbf{F}$ down with a right angle onto the x axis, and that gives the component of the force in that

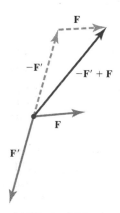

FIGURE 1-4   Vector subtraction, first method.

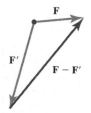

FIGURE 1-5   Vector subtraction, second method.

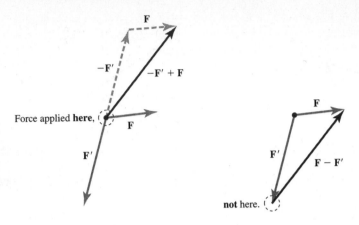

FIGURE 1-6 Subtraction of two forces applied at the same point.

direction, which we call F_x. Mathematically, F_x is the *magnitude* of **F** (which I'll write $|\mathbf{F}|$) times the cosine of the angle that **F** makes with the x axis; this comes from the properties of the right triangle. (See Fig. 1-7.)

$$F_x = |\mathbf{F}| \cos \theta. \tag{1.10}$$

Now, if **A** and **B** are added to make **C**, then the projections that are brought down to form a right angle in a given direction 'x', evidently add. So the components of the vector sum are the sum of the vector components, and that's true of components *in any direction*. (See Fig. 1-8.)

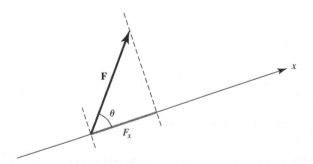

FIGURE 1-7 The component of vector **F** in direction x.

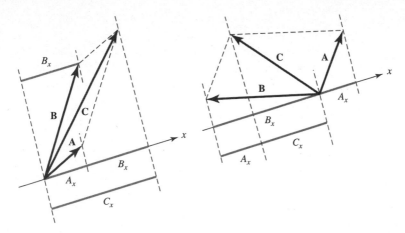

FIGURE 1-8 A component of a vector sum equals the sum of the corresponding vector components.

$$\mathbf{A} + \mathbf{B} = \mathbf{C} \Rightarrow A_x + B_x = C_x. \qquad (1.11)$$

Particularly convenient is the description of vectors in terms of their components on perpendicular axes, x and y (and z—there's three dimensions in the world; I keep forgetting that, because I'm always drawing on a blackboard!). If we have a vector $\mathbf{F}$ that is in the x-y plane, and we know its component in the x direction, that doesn't completely define $\mathbf{F}$, because there are many vectors in the x-y plane that have the same component in the x direction. But if we also know $\mathbf{F}$'s component in the y direction, then $\mathbf{F}$ is completely specified. (See Fig. 1-9.)

The components of $\mathbf{F}$ along the x, y, and z axes can be written as F_x, F_y, and F_z; summing vectors is equivalent to summing their components, so if the components of another vector $\mathbf{F}'$ are F_x', F_y', and F_z', then $\mathbf{F} + \mathbf{F}'$ has the components $F_x + F_x'$, $F_y + F_y'$, and $F_z + F_z'$.

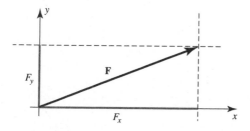

FIGURE 1-9 A vector in the x-y plane is completely specified by two components.

That's the easy part; now it gets a bit more difficult. There's a way of multiplying two vectors to produce a *scalar*—a number that is the same in any coordinate system. (In fact, there's a way of making a scalar out of one vector, and I'll come back to that.) You see, if the coordinate axes change, then the components change—but the angle between vectors and their magnitudes stay the same. If **A** and **B** are vectors, and the angle between them is θ, I can take the magnitude of **A**, times the magnitude of **B** times the cosine of θ, and call this number **A** · **B** ("**A** *dot* **B**"). (See Fig. 1-10.) That number, called a "dot product" or a "scalar product," is the same in all coordinate systems:

$$\mathbf{A} \cdot \mathbf{B} = |\mathbf{A}||\mathbf{B}| \cos \theta. \qquad (1.12)$$

It is evident that since $|\mathbf{A}| \cos \theta$ is the projection of **A** onto **B**, **A** · **B** is equal to the projection of **A** onto **B** times the magnitude of **B**. Similarly, since $|\mathbf{B}| \cos \theta$ is the projection of **B** onto **A**, **A** · **B** also equals the projection of **B** onto **A** times the magnitude of **A**. However, I find for myself that $\mathbf{A} \cdot \mathbf{B} = |\mathbf{A}||\mathbf{B}| \cos \theta$ is the easiest way to remember what the dot product is; then I can always see the other relations immediately. The trouble is, of course, you have so many ways of saying the same thing that it's no good to try to remember them all—a point that I'll make, in a few minutes, more completely.

We can also define **A** · **B** in terms of the components of **A** and **B** on an arbitrary set of axes. If I were to take three mutually perpendicular axes, x, y, z, in some arbitrary orientation, then **A** · **B** will turn out to be

$$\mathbf{A} \cdot \mathbf{B} = A_x B_x + A_y B_y + A_z B_z. \qquad (1.13)$$

It is not immediately self-evident how you get from $|\mathbf{A}||\mathbf{B}| \cos \theta$ to $A_x B_x + A_y B_y + A_z B_z$. Although I can prove it when I want to,[3] it takes me too long, so I remember them both.

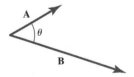

FIGURE 1-10 The vector dot product $|\mathbf{A}||\mathbf{B}| \cos \theta$ is the same in all coordinate systems.

---

[3] See *The Feynman Lectures on Physics (FLP)* Vol. I, Section 11-7.

When we take the dot product of a vector with *itself,* θ is 0, and the cosine of 0 is 1, so $\mathbf{A} \cdot \mathbf{A} = |\mathbf{A}||\mathbf{A}| \cos 0 = |\mathbf{A}|^2$. In terms of components, it's $\mathbf{A} \cdot \mathbf{A} = A_x^2 + A_y^2 + A_z^2$. The positive square root of that number is the magnitude of the vector.

1-7 Differentiating vectors

Now, we can do what's called differentiating the vectors. The derivative of a vector with respect to time is meaningless unless the vector depends on the time, of course. That means we have to imagine some vector that is different all the time: as time goes on, the vector keeps changing, and we want the rate of change.

For example, the vector $\mathbf{A}(t)$ might be the position, at time t, of an object that's flying around. At the next moment, t', the object has moved from $\mathbf{A}(t)$ to $\mathbf{A}(t')$; we would like to calculate the rate of change of $\mathbf{A}$ at time t.

The rule is the following: that in the interval $\Delta t = t' - t$, the thing has moved from $\mathbf{A}(t)$ to $\mathbf{A}(t')$, so the displacement is $\Delta\mathbf{A} = \mathbf{A}(t') - \mathbf{A}(t)$, a difference vector from the old position to the new position. (See Fig. 1-11.)

Of course, the shorter the interval Δt, the closer $\mathbf{A}(t')$ is to $\mathbf{A}(t)$. If you divide $\Delta\mathbf{A}$ by Δt and then take the limit as they both approach zero—that's the derivative. In this case, where $\mathbf{A}$ is position, its derivative is a velocity vector; the velocity vector is in a direction tangent to the curve, because that's the direction of the displacements; its magnitude you can't get by looking at this picture, because it depends on how *fast* the thing is going along the curve. The magnitude of the velocity vector is the speed; it tells you how far the thing moves per unit time. So, that's a definition of the velocity vector: it's tangent to the path, and its magnitude is equal to the speed of motion on the path. (See Fig. 1-12.)

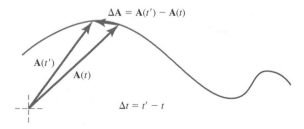

FIGURE 1-11   Position vector **A** and displacement Δ**A** during interval Δ*t*.

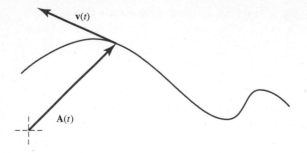

FIGURE 1-12 Position vector **A** and its derivative **v** at time *t*.

$$\mathbf{v}(t) = \frac{d\mathbf{A}}{dt} = \lim_{\Delta t \to 0} \frac{\Delta \mathbf{A}}{\Delta t}. \tag{1.14}$$

Incidentally, it is dangerous to draw both the position vector and the velocity vector in the same diagram, unless you're being very careful—and since we're having a little trouble understanding these things, I point out all the possible pitfalls that I can think of, because the next thing you might want to do is *add* **A** to **v** for some purpose. That's not legitimate, because in order to really draw the velocity vector, you have to know the scale of time: the velocity vector is in a different scale than the position vector; in fact, they have different units. You can't add positions and velocities together in general—and you can't add them here.

In order for me to actually *draw* the picture of any vector, I have to make a decision as to the scale. When we talked about forces, we said that so-and-so many newtons were going to be represented by 1 inch (or 1 meter, or whatever). And here, we have to say that so-and-so many meters per second is going to be represented by 1 inch. Someone else could draw the picture with position vectors the same lengths as ours, but with the velocity vector one-third as long as ours—he's just using a different scale for his velocity vector. There's no unique way to draw the length of a vector because the choice of scale is arbitrary.

Now, the velocity in terms of *x*, *y*, and *z* components is very easy, because, for example, the rate of change of the *x* component of position is equal to the *x* component of velocity, and so on. This is simply because the derivative is really a difference, and since the components of a difference vector equal the differences of the corresponding components, we have

$$\left(\frac{\Delta \mathbf{A}}{\Delta t}\right)_x = \frac{\Delta A_x}{\Delta t}, \quad \left(\frac{\Delta \mathbf{A}}{\Delta t}\right)_y = \frac{\Delta A_y}{\Delta t}, \quad \left(\frac{\Delta \mathbf{A}}{\Delta t}\right)_z = \frac{\Delta A_z}{\Delta t}, \tag{1.15}$$

and then taking limits we have the components of the derivative:

$$v_x = \frac{dA_x}{dt}, \quad v_y = \frac{dA_y}{dt}, \quad v_z = \frac{dA_z}{dt}. \tag{1.16}$$

This is true for any direction: if I take the component of $\mathbf{A}(t)$ in any direction, then the velocity vector component in that direction is the derivative of the component of $\mathbf{A}(t)$ in that direction, with one serious warning: the direction must not change with time. You can't say, "I'm gonna take the component of $\mathbf{A}$ in the direction of $\mathbf{v}$," or something like that, because $\mathbf{v}$ is *moving*. It's only true that the derivative of the position component is equal to the velocity component *if the direction in which you take the component is itself fixed.* So equations (1.15) and (1.16) are only true for x, y, z, and other fixed axes; if the axes are turning while you're trying to take the derivative, the formula is much more complicated.

Those are some of the deviations and difficulties of differentiating vectors.

Of course, you can differentiate the derivative of a vector, then differentiate that, and so on. I called the derivative of $\mathbf{A}$ "velocity," but that's only because $\mathbf{A}$ is the position; if $\mathbf{A}$ is something else, its derivative is something other than velocity. For example, if $\mathbf{A}$ is the momentum, the time derivative of momentum equals the force, so the derivative of $\mathbf{A}$ would be the force. And if $\mathbf{A}$ were the velocity, the time derivative of the velocity is the acceleration, and so on. What I've been telling you is generally true of differentiating vectors, but here I've given only the example of positions and velocities.

1-8 Line integrals

Finally, there's only one more thing that I have to talk about for vectors, and that is a horrible, complicated thing, called a "line integral":

$$\int_a^z \mathbf{F} \cdot d\mathbf{s}. \tag{1.17}$$

We'll take as an example that you have a certain vector field $\mathbf{F}$, which you want to integrate along a curve S from point a to point z. Now, in order for this line integral to mean something, there must be some way of defining the value of $\mathbf{F}$ at every point on S between a and z. If $\mathbf{F}$ is defined as the force applied to an object at point a, but you can't tell me how the force changes as you move along S, *at least* between a and z, then "the integral

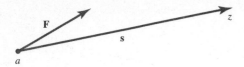

FIGURE 1-13 A constant force **F** defined on the straight-line path *a-z*.

of **F** along S from a to z" makes no *sense*. (I said "at least," because **F** could be defined anywhere else too, but at least you must define it on the part of the curve that you are integrating along.)

In a moment I'll define the line integral of an arbitrary vector field along an arbitrary curve, but first let's consider the case where **F** is constant, and S is a straight-line path from a to z—a displacement vector, which I'll call **s**. (See Fig. 1-13.) Then, since **F** is constant, we can take it outside the integral (just like ordinary integration), and the integral of d**s** from a to z is just **s**, so the answer is **F** · **s**. That's the line integral for a constant force and a straight-line path—the easy case:

$$\int_a^z \mathbf{F} \cdot d\mathbf{s} = \mathbf{F} \cdot \int_a^z d\mathbf{s} = \mathbf{F} \cdot \mathbf{s}. \qquad (1.18)$$

(Remember that **F** · **s** is the component of the force in the direction of the displacement times the magnitude of the displacement; in other words, it's simply the distance along the line times the component of force in that direction. There are a lot of other ways to look at it, too: it's the component of the displacement in the direction of the force, times the magnitude of the force; it's the magnitude of the force times the magnitude of the displacement, times the cosine of the angle between them. These are all equivalent.)

More generally, the line integral is defined as follows. First, we break up the integral by dividing S between a and z into N equal segments: ΔS_1, $\Delta S_2 \ldots \Delta S_N$. Then the integral along S is the integral along ΔS_1 plus the integral along ΔS_2 plus the integral along ΔS_3, and so on. We choose N large so that we can approximate each ΔS_i by a little displacement vector, $\Delta \mathbf{s}_i$, over which **F** has an approximately constant value, $\mathbf{F}_i$. (See Fig. 1-14.) Then, by the "constant force straight-line path" rule, segment ΔS_i contributes approximately $\mathbf{F}_i \cdot \Delta \mathbf{s}_i$ to the integral. So, if you add together $\mathbf{F}_i \cdot \Delta \mathbf{s}_i$ for i equals 1 to N, that's an excellent approximation to the integral. The integral is *exactly* equal to this sum only if we take the limit as N goes

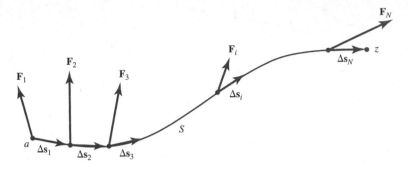

FIGURE 1-14 A variable force **F** defined on the curve S.

to infinity: you take the segments as fine as you can; you take them a little finer than that, and you get the correct integral:

$$\int_a^z \mathbf{F} \cdot d\mathbf{s} = \lim_{N \to \infty} \sum_{i=1}^{N} \mathbf{F}_i \cdot \Delta\mathbf{s}_i. \tag{1.19}$$

(This integral, of course, depends upon the curve—generally—though sometimes it doesn't in the physics.)

Well, then, that's all there is to the mathematics that you have to know to do the physics—for now, at least—and these things, most particularly the calculus and the early parts of the vector theory, should become second nature. Some things—like the line integral—may not be second nature *now,* but they will be, eventually, as you use them more; they aren't so vital *yet,* and that's harder. The things you "gotta get into your head good," right now, are the calculus, and the little things about taking the components of vectors in various directions.

1-9   A simple example

I'll give one example—just a very simple one—to show how to take components of vectors. Suppose we have a machine of some kind, as illustrated in Figure 1-15: it's got two rods connected by a pivot (like an elbow joint) with a big weight on it. The end of one rod is connected to the floor by a stationary pivot, and the end of the other rod has a rolling pivot that rolls along the floor in a slot—it's part of a machine, see, and it's going choo-choog, choo-choog—the roller's going back and forth, the weight's going up and down, and so on.

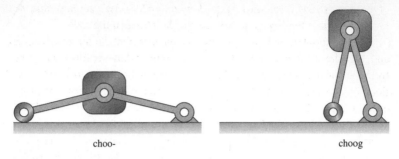

choo-   choog

FIGURE 1-15   A simple machine.

Let's say the weight is 2 kg, the rods are 0.5 meters long, and at a certain moment when the machine is standing still, the distance from the weight to the floor just happens to come out, luckily, to 0.4 meters—so that we have a 3-4-5 triangle, to make the arithmetic easier. (See Fig. 1-16.) (The arithmetic shouldn't make any difference; the real difficulty is to get the *ideas* right.)

The problem is to figure out what horizontal push **P** you have to make on the roller in order to *hold* that weight up. Now, I'm going to make an assumption that we will need in order to do the problem. We make the assumption that when a rod has pivots at *both* ends, then the net force is always directed *along the rod*. (It turns out to be true; you may feel it's self-evident.) It would not necessarily be true if there were a pivot only at *one* end of the rod, because then I could push the rod sideways. But if there's a

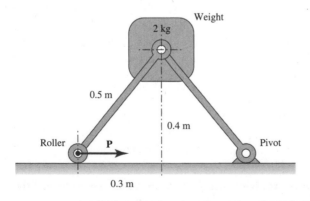

FIGURE 1-16   What force, **P**, is required to hold up the weight?

pivot at both ends, I can only push *along* the rod. So let's suppose that we know that—that the forces must lie in the directions of the rods.

We also know something else from the physics: that the forces are equal and opposite at the ends of the rods. For example, whatever force is exerted by the rod on the roller must also be exerted by that rod, in the opposite direction, on the weight. So, that's the problem: with these ideas about the properties of rods, we try to figure out what's the horizontal force on the roller.

I think the way I'd like to try to do it is this: the horizontal force exerted on the roller by the rod is a certain component of the net force on it. (Of course, there's also a vertical component due to the "confining slot," which is unknown and uninteresting; it's part of the net force on the roller, which is exactly opposite the net force on the weight.) Therefore I can get the components of the force exerted on the roller by the rod—in particular, the horizontal component I want—if I can get the components of the force exerted by the rod on the weight. If I call the horizontal force on the weight F_x, then the horizontal force on the roller is $-F_x$, and the force needed to hold the weight up is equal and opposite to *that*, so $|\mathbf{P}| = F_x$.

The vertical force on the weight from the rod, F_y, is very easy: it's simply equal to the weight of the thing, which is 2 kg, times g, the gravitational constant. (Something else you have to know from physics—g is 9.8, in the mks system.) F_y is 2 times g, or 19.6 newtons, so the vertical force on the roller is –19.6 newtons. Now, how can I get the horizontal force? Answer: I get it by knowing that the net force must lie along the rod. If F_y is 19.6, and the net force lies along the rod, then how much must F_x be? (See Fig. 1-17.)

Well, we have the projections of the triangles, which have been designed very nicely, so that the ratio of the horizontal to the vertical sides is 3 to 4; that's the same ratio as F_x is to F_y, (I don't care about the net force, $\mathbf{F}$, here; I only need the force in the *horizontal* direction) and I already know what the vertical force is. So, the magnitude of the horizontal force—unknown—is to 19.6 as 0.3 is to 0.4. Therefore I multiply 3/4 by 19.6 and I get:

$$\frac{F_x}{19.6} = \frac{0.3}{0.4}.$$

$$\therefore F_x = \frac{0.3}{0.4} \times 19.6 = 14.7 \text{ newtons.}$$

(1.20)

We conclude that $|\mathbf{P}|$ the horizontal force on the roller needed to hold the weight up, is 14.7 newtons. That's the answer to this problem.

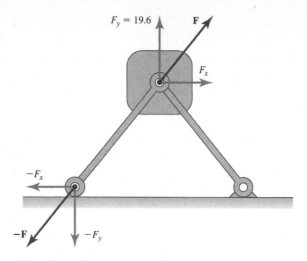

FIGURE 1-17 The force on the weight and the force on the roller from one rod.

Or *is* it?

You see, you can't do physics just by plugging in the formulas: you'll never get anywhere without having something else besides knowing the rules, the formulas for projections, and all that stuff; you have to have a certain *feeling* for the real situation! I'll make some more remarks about that in a minute, but here, in this particular problem, the difficulty is the following: the net force on the weight is not only from *one* rod, there's also a force exerted on it by the *other* rod, in some direction, and I left that out when I made the analysis—so it's all wrong!

I also have to worry about the force that the rod with the stationary pivot exerts on the weight. Now it's getting complicated: how can I figure out what *that* force is? Well, what is the net force of *everything* on the weight? Just the gravity—it just balances the gravity; there is no force horizontally on the weight. So the *clue* by which I can find out how much "juice" there is along the rod with the stationary pivot, is to notice that it must exert just enough horizontally to balance the horizontal force that the other rod is exerting.

Therefore, if I were to draw the force that the rod with the stationary pivot exerts, its horizontal component would be exactly opposite the horizontal component that the rod with the roller exerts, and the vertical components would be equal because of the identical 3-4-5 triangles the rods make: both rods are pushing up the same amount because their horizontal components must balance—if the rods were different lengths, you'd have a little more work to do, but it's the same idea.

So, let's start out with the weight again: the forces *from the rods on the weight* are the first things to get straightened out. So, let's look at the forces *from the rods on the weight.* The reason I keep repeating this to myself is because otherwise I get the signs all mixed up: The force *from the weight on the rods* is the opposite of the force *from the rods on the weight.* I always have to start over after I get all balled up like this; I have to think it out again, and make up my mind as to what I want to talk about. So I say, "Look at the forces *from the rods on the weight:* there's a force **F**, which is in the direction of one rod. Then there's a force **F′**, in the direction of the other rod. Those are the only two forces, and they are in the directions of the rods."

Now, the net of these two forces—ahhhh! I'm beginning to see the light! The *net* of these two forces has *no* horizontal component, and a vertical component of 19.6 newtons. Ah! Let me draw the picture again, since I did it wrong before. (See Fig. 1-18.)

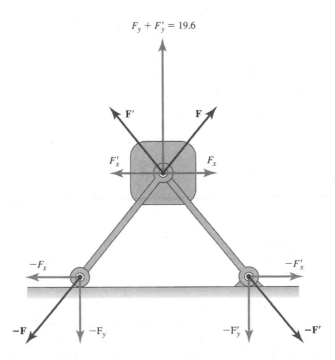

FIGURE 1-18 The force on the weight and the forces on the roller and pivot, from both rods.

The horizontal forces balance, therefore the vertical components add, and the 19.6 newtons is not just the vertical component of the force from *one* rod, but the total from both; since each rod contributes half, the vertical component from the rod with the roller is only 9.8 newtons.

Now when we take the horizontal projection of this force, multiplying it by 3/4 as we did before, we get the horizontal component of force from the rod with the roller on the weight, and that takes care of that:

$$\frac{F_x}{9.8} = \frac{0.3}{0.4}. \tag{1.21}$$

$$\therefore \ F_x = \frac{0.3}{0.4} \times 9.8 = 7.35 \text{ newtons.}$$

1-10   Triangulation

I have a few moments left, so I'd like to make a little speech about the relation of the mathematics to the physics—which, in fact, was well illustrated by this little example. It will not do to memorize the formulas, and to say to yourself, "I know all the formulas; all I gotta do is figure out how to put 'em in the problem!"

Now, you may succeed with this for a while, and the more you work on memorizing the formulas, the longer you'll go on with this method—but it doesn't work in the end.

You might say, "I'm not gonna believe him, because I've always been successful: that's the way I've always done it; I'm always gonna do it that way."

You are *not* always going to do it that way: you're going to *flunk*—not this year, not next year, but eventually, when you get your job, or something—you're going to lose along the line somewhere, because physics is an *enormously* extended thing: there are *millions* of formulas! It's impossible to remember all the formulas—it's *impossible!*

And the great thing that you're ignoring, the powerful machine that you're not using, is this: suppose Figure 1-19 is a map of all the physics formulas, all the relations in physics. (It should have more than two dimensions, but let's suppose it's like that.)

Now, suppose that something happened to your mind, that somehow all the material in some region was erased, and there was a little spot of missing goo in there. The relations of nature are so nice that it is possible, by logic, to "triangulate" from what is known to what's in the hole. (See Fig. 1-20.)

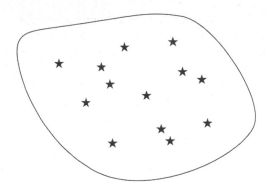

FIGURE 1-19 Imaginary map of all the physics formulas.

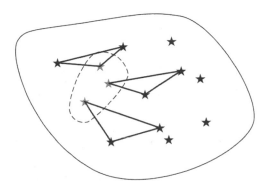

FIGURE 1-20 Forgotten facts can be recreated by triangulating from known facts.

And you can re-create the things that you've forgotten *perpetually*—if you don't forget too much, and if you know enough. In other words, there comes a time—which you haven't quite got to, yet—where you'll know so many things that as you forget them, you can reconstruct them from the pieces that you can still remember. It is therefore of first-rate importance that you know how to *"triangulate"*—that is, to know how to figure something out from what you already know. *It is absolutely necessary.* You might say, "Ah, I don't care; I'm a *good* memorizer! I know how to *really* memorize! In fact, I took a *course* in memory!"

That *still* doesn't work! Because the real utility of physicists—both to discover new laws of nature, and to develop new things in industry, and so

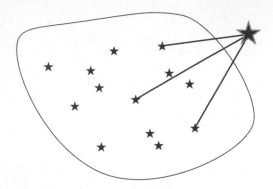

FIGURE 1-21 New discoveries are made by physicists triangulating from the known to the previously unknown.

on—is *not* to talk about what's already known, but to do something *new*—and so they triangulate out from the known things: they make a "triangulation" that *no one has ever made before.* (See Fig. 1-21.)

In order to learn how to do that, you've got to forget the memorizing of formulas, and to try to learn to *understand the interrelationships* of nature. That's very much more difficult at the beginning, but it's the *only successful way.*

2 *Laws and Intuition*

REVIEW LECTURE B

Last time we discussed the mathematics that you need to know to do the physics, and I pointed out that equations should be memorized as a tool, but that it isn't a good idea to memorize everything. In fact, it's impossible in the long run to do everything by memory. That doesn't mean to do *nothing* by memory—the more you remember, in a certain sense, the better it is—but you should be able to re-create anything that you forgot.

Incidentally, on the subject of suddenly finding yourself below average when you come to Caltech, which we also discussed last time, if you somehow escape from being in the bottom half of the class, you're just making it miserable for somebody else, because now you're forcing somebody *else* to go down to the bottom half! But there *is* a way you can do it without disturbing anybody: find and pursue something interesting that delights you especially, so you become a kind of temporary expert in some phenomenon that you heard about. It's the way to save your soul—then you can always say, "Well, at least the other guys don't know anything about *this!*"

2-1 The physical laws

Now, in this review, I'm going to talk about the physical laws, and the first thing to do is to state what they are. We stated them in words a lot during the lectures so far, and it's hard to say it all again without using the same amount of time, but the physical laws can also be summarized by some equations, which I'll write down here. (By this time I'll suppose that your mathematics is developed to a point that you can understand the notation right away.) The following are all the physical laws that you should know.

First:

$$\mathbf{F} = \frac{d\mathbf{p}}{dt}. \tag{2.1}$$

That is, the force, $\mathbf{F}$, is equal to the rate of change, with respect to time, of the momentum, $\mathbf{p}$. ($\mathbf{F}$ and $\mathbf{p}$ are vectors. You're supposed to know what the symbols mean by this time.)

I'd like to emphasize that in any *physical* equation it is necessary to understand what the letters stand for. That doesn't mean to say, "Oh, I know that's **p**, which stands for the mass in motion times the velocity, or the mass at rest times the velocity over the square root of 1 minus v squared over c squared":[1]

$$\mathbf{p} = \frac{m\mathbf{v}}{\sqrt{1 - v^2/c^2}}. \tag{2.2}$$

Instead, to understand *physically* what the **p** stands for, you have to know that **p** is not just "the momentum"; it's the momentum of *something*—the momentum of a *particle* whose mass is m and whose velocity is **v**. And, in Eq. (2.1), **F** is the total force—the vector sum of all the forces that are acting on that particle. Only *then* can you have an understanding of these equations.

Now, here's another physical law that you should know, called the conservation of momentum:

$$\sum_{\text{particles}} \mathbf{P}_{after} = \sum_{\text{particles}} \mathbf{P}_{before}. \tag{2.3}$$

The law of conservation of momentum says that the total momentum is a constant in any situation. What does that mean, physically? For instance in a collision, it's the same as saying that the sum of the momenta of all the particles *before* a collision is the same as the sum of the momenta of all the particles *after* the collision. In the relativistic world, the particles can be different after the collision—you can create new particles and destroy old ones—but it's still true that the vector sum of the total momenta of everything before and after is the same.

The next physical law you should know, called the conservation of energy, takes the same form:

$$\sum_{\text{particles}} E_{after} = \sum_{\text{particles}} E_{before}. \tag{2.4}$$

That is, the sum of the energies of all the particles *before* a collision is equal to the sum of the energies of all the particles *after* the collision. In order to use this formula, you have to know what the energy of a particle is. The energy of a particle with rest mass m and speed v is

$$E = \frac{mc^2}{\sqrt{1 - v^2/c^2}}. \tag{2.5}$$

---

[1] $v = |\mathbf{v}|$ is the speed of the particle; c is the speed of light.

2-2 The nonrelativistic approximation

Now, those are the laws that are correct in the relativistic world. In the *non-relativistic approximation*—that is, if we look at particles at *low* velocity compared to the speed of light—then there are some special cases of the above laws.

To begin with, the momentum at low velocities is easy: $\sqrt{1 - v^2/c^2}$ is almost 1, so Eq. (2.2) becomes

$$\mathbf{p} = m\mathbf{v}. \tag{2.6}$$

That means the formula for the force, $\mathbf{F} = d\mathbf{p}/dt$, can also be written $\mathbf{F} = d(m\mathbf{v})/dt$. Then, by moving the constant, m, out in front, we see that for low velocities, the force equals the mass times the acceleration:

$$\mathbf{F} = \frac{d\mathbf{p}}{dt} = \frac{d(m\mathbf{v})}{dt} = m\frac{d\mathbf{v}}{dt} = m\mathbf{a}. \tag{2.7}$$

The conservation of momentum for particles at low velocities has the same form as Eq. (2.3), except that the formula for the momenta is $\mathbf{p} = m\mathbf{v}$ (and the masses are all constant):

$$\sum_{particles} (m\mathbf{v})_{after} = \sum_{particles} (m\mathbf{v})_{before}. \tag{2.8}$$

However, the conservation of *energy* at low velocities becomes *two* laws: first, that the *mass* of *each particle* is constant—you can't create or destroy any material—and second, that the sum of the $\frac{1}{2}mv^2$s (the total kinetic energy, or *K.E.*) of all the particles is constant:[2]

$$m_{after} = m_{before}$$

$$\sum_{particles} \left(\tfrac{1}{2}mv^2\right)_{after} = \sum_{particles} \left(\tfrac{1}{2}mv^2\right)_{before}. \tag{2.9}$$

---

[2]The relationship between the kinetic energy of a particle and its total (relativistic) energy can readily be seen by substituting the first two terms of the Taylor series expansion of $1/\sqrt{1 - v^2/c^2}$ into Eq. (2.5):

$$\frac{1}{\sqrt{1 - x^2}} = 1 + \frac{1}{2}x^2 + \frac{1 \cdot 3}{2 \cdot 4}x^4 + \frac{1 \cdot 3 \cdot 5}{2 \cdot 4 \cdot 6}x^6 + \dots$$

$$E = \frac{mc^2}{\sqrt{1 - v^2/c^2}} = mc^2(1 + v^2/2c^2 + \cdots)$$

$$\approx mc^2 + \tfrac{1}{2}mv^2 = rest\ energy + K.E. \quad (\text{for } v \ll c).$$

If we think of large, everyday objects as particles with low velocities—
like an ashtray is a particle, approximately—then the law that the sum of
the kinetic energies before equals the sum after is *not* true, because there
can be some $\frac{1}{2}mv^2$s of the particles all mixed up on the inside of the objects,
in the form of internal motion—heat, for example. So in a collision between
large objects, this law appears to fail. It's only true for fundamental particles. Of course with large objects, it can happen that *not much* energy goes
into the internal motion, so the conservation of energy appears to be *nearly*
true, and that's called a *nearly elastic* collision—which is sometimes ideal-
ized as a *perfectly elastic* collision. So energy is much more difficult to
keep track of than momentum, because the conservation of kinetic energy
needn't be true when the objects involved are large, like weights and so on,
that have inelastic collisions.

2-3   Motion with forces

Now, if we look not at a collision, but at motion when forces act—then we
get first a theorem that tells us that the *change in kinetic energy* of a particle is equal to the *work* done on it by the forces:

$$\Delta K.E. = \Delta W. \qquad (2.10)$$

Remember, this *means* something—you have to know what all the letters mean: it means that if a particle is moving on some curve, S, from
point A to point B, and it's moving under the influence of a force $\mathbf{F}$, where
$\mathbf{F}$ is the total force acting on the particle, then if you knew what the $\frac{1}{2}mv^2$
of the particle is at point A, and what it is over at point B, they differ by the
integral, from A to B, of $\mathbf{F} \cdot d\mathbf{s}$, where $d\mathbf{s}$ is an increment of displacement
along S. (See Fig. 2-1).

$$\Delta K.E. = \tfrac{1}{2}mv_B^2 - \tfrac{1}{2}mv_A^2 \qquad (2.11)$$

and

$$\Delta W = \int_A^B \mathbf{F} \cdot d\mathbf{s}. \qquad (2.12)$$

In certain cases, that integral can be calculated easily ahead of time,
because the force on the particle depends only on its position in a simple
way. Under those circumstances we can write that the work done on the particle is equal in magnitude and opposite to the change in another quantity
called its *potential energy, or P.E.* Such forces are said to be "conservative":

$$\Delta W = -\Delta P.E. \quad \text{(with a } conservative \text{ force, } \mathbf{F}). \qquad (2.13)$$

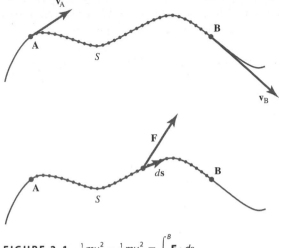

FIGURE 2-1 $\frac{1}{2}mv_B^2 - \frac{1}{2}mv_A^2 = \int_A^B \mathbf{F} \cdot d\mathbf{s}.$

Incidentally, the words that we use in physics are terrible: "conservative forces" doesn't mean that the *forces* are conserved, but rather that the forces are such that the *energy* of the things that the forces work on *can be* conserved.[3] It's very confusing, I admit, and I can't help it.

The total energy of a particle is its kinetic energy plus its potential energy:

$$E = K.E. + P.E. \tag{2.14}$$

When only conservative forces act, a particle's total energy does not change:

$$\Delta E = \Delta K.E. + \Delta P.E. = 0 \quad \text{(with \textit{conservative} forces).} \tag{2.15}$$

But when *nonconservative* forces act—forces not included in any potential—then the change in a particle's energy is equal to the work done on it by those forces.

$$\Delta E = \Delta W \quad \text{(with \textit{nonconservative} forces).} \tag{2.16}$$

---

[3]A force is defined to be conservative when the total work it does on a particle that moves from one place to another is the same regardless of the path the particle moves on—the total work done depends only on the endpoints of the path. In particular, the work done by a conservative force on a particle that goes around a closed path, ending where it began, is always zero. See *FLP* Vol. I, Section 14-3.

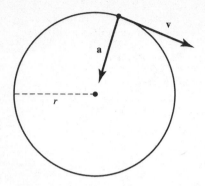

FIGURE 2-2 Velocity and acceleration vectors for constant-speed circular motion.

Now, the end of this part of the review comes when we give all the rules that are known for the various forces.

But before I do that, there's a formula for acceleration that is very useful: if, at a given instant, a thing is moving on a circle of radius r at constant speed v, then its acceleration is directed towards the center, and is equal in magnitude to v^2/r. (See Fig. 2-2.) That's sort of at "right angles" to everything else I've been talking about, but it's good to remember that formula, because it's a pain in the neck to derive it:[4]

$$|\mathbf{a}| = \frac{v^2}{r}. \tag{2.17}$$

TABLE 2-1

	True always	False in general (true only at low velocities)
Force	$\mathbf{F} = \dfrac{d\mathbf{p}}{dt}$	$\mathbf{F} = m\mathbf{a}$
Momentum	$\mathbf{p} = \dfrac{m\mathbf{v}}{\sqrt{1-v^2/c^2}}$	$\mathbf{p} = m\mathbf{v}$
Energy	$E = \dfrac{mc^2}{\sqrt{1-v^2/c^2}}$	$E = \frac{1}{2}mv^2 \left(+mc^2\right)$

[4]See *FLP* Vol. I, Section 11-6.

TABLE 2-2

True with conservative forces	True with nonconservative forces
$\Delta P.E. = -\Delta W$	P.E. is undefined.
$\Delta E = \Delta K.E. + \Delta P.E. = 0$	$\Delta E = \Delta W$

Definitions: Kinetic Energy, $K.E. = \frac{1}{2}mv^2$; Work, $W = \int \mathbf{F} \cdot d\mathbf{s}$.

2-4 Forces and their potentials

Now, to get back on the track, I will list a series of laws of force, and the formulas for their potentials.

TABLE 2-3

	Force	Potential
Gravity, near the earth's surface	$-mg$	mgz
Gravity, between particles	$-Gm_1m_2/r^2$	$-Gm_1m_2/r$
Electric Charge	$q_1q_2/4\pi\epsilon_0 r^2$	$q_1q_2/4\pi\epsilon_0 r$
Electric Field	$q\mathbf{E}$	$q\phi$
Ideal Spring	$-kx$	$\frac{1}{2}kx^2$
Friction	$-\mu N$	No!

First is surface gravity on the earth. The force is down, but never mind the sign; just remember which direction the force is, because who knows what your axes are—maybe you're making the z axis down! (You're allowed to.) So the force is $-mg$, and potential energy is mgz, where m is the mass of an object, g is a constant (the acceleration of gravity at the surface of the earth—otherwise, the formula is no good!), and z is the height above the ground, or any other level. That means the value of the potential energy can be zero any place you want. The way we're going to use potential energy is to talk about its *changes*—and then, of course, it doesn't make any difference if you add a constant.

Next is gravity in space between particles; this force is directed centrally, and is proportional to the product of the one mass by the other mass divided by the distance between the two squared, $-mm'/r^2$, or $-m_1m_2/r^2$, or any other way you want to write it. It's better to just remember which direction the force is, than to worry about the sign. But this part you've got to remember: the force of gravity goes as the inverse square of the distance between the particles. (So which way *is* the sign? Well, likes attract in gravity, so the force is in the opposite direction to the radius vector. That shows you that I don't remember the sign; I just remember *physically* which way the sign is: the particles *attract*—that's all I have to remember.)

Now, the *potential* energy between two particles is $-Gm_1m_2/r$. It's hard for me to remember which way the potential energy goes. Let's see: the particles lose potential energy when they come together, so that means when r is smaller, the potential energy should be less, so it's negative—I *hope* that's right! I have a great deal of difficulty with signs.

For electricity, the force is proportional to the product of the charges, q_1 and q_2, divided by the distance between them squared. But the constant of proportionality, instead of being written in the numerator (as with gravity), is written as $4\pi\epsilon_0$ in the denominator. The electrical force is directed radially, just like gravitation is, but with the opposite law of sign: likes *repel,* electrically, and therefore the sign of electrical potential energy is opposite that of gravitational potential energy, but then the constant of proportionality is different: $1/4\pi\epsilon_0$ instead of G.

Some technical points from the laws of electricity: the force on q units of charge can be written as q times the electric field, $q\mathbf{E}$, and the energy can be written as q times the electrical potential, $q\phi$. Here, $\mathbf{E}$ is a vector field and ϕ is a scalar field. q is measured in *coulombs,* and ϕ is measured in *volts*—when the energy is in the usual units of *joules.*

To continue this table of formulas, we have next an ideal spring. The force to pull out an ideal spring to a distance x is a constant, k, times x. Now, you have to know what the letters mean again: x is the distance that you pull the spring away from the equilibrium position, and the force pulls it back an amount $-kx$. I put the sign in just to say the spring pulls backwards; you know damn well a spring pulls a thing *back,* and doesn't push it out further when you pull on it. Now, the potential energy is $\frac{1}{2}kx^2$. In order to pull out a spring you do work on it, so after it's pulled out, the potential energy is plus. So this sign business is easy—for the spring.

You see, details like the signs that I can't remember, I try to reconstruct by arguments—that's how I remember all the things I don't remember.

Friction: the force of friction against a dry surface is $-\mu N$, and again you have to know what the symbols mean: when an object is pushed against another surface with a force whose component perpendicular to the surface is N, then in order to keep it sliding along the surface, the force required is

μ times N. You can easily figure out which direction the force is; it's oppo-site to the direction you slide it.

Now, under the potential energy for friction in Table 2-3, the answer is *No:* friction does not conserve energy, and therefore we have no formula for the potential energy for friction. If you push an object along a surface one way, you do work; then, when you drag it back, you do work again. So after you've gone through a complete cycle, you haven't come out with *no* energy change; you've done work—and so friction has no potential energy.

2-5 Learning physics by example

Those are all the rules I can remember as being necessary. So you say, "Well, that's very easy: I'll just memorize the whole damn table, and then I'll know all the physics." Well, it won't work.

Actually, it might work fairly well at the beginning, but it gets harder and harder, as I pointed out in Chapter 1. Therefore, what we have to learn next is how to apply the mathematics to the physics in order to understand the *world.* The equations keep track of things for us, so we use them as tools—but to do that, we have to know what *objects* the equations are talking about.

The problem of how to deduce new things from old, and how to solve problems, is really very difficult to teach, and I don't really know how to do it. I don't know how to tell you something that will transform you from a person who *can't* analyze new situations or solve problems, to a person who *can.* In the case of the mathematics, I can transform you from some-body who *can't* differentiate to somebody who *can,* by giving you all the rules. But in the case of the physics, I can't transform you from somebody who *can't* to somebody who *can,* so I don't know what to do.

Because I *intuitively* understand what's going on physically, I find it dif-ficult to communicate: I can only do it by showing you examples. Therefore, the rest of this lecture, as well as the next one, will consist of doing a whole lot of little examples—of applications, of phenomena in the physical world or in the industrial world, of applications of physics in dif-ferent places—to show you how what you already know will permit you to understand or to analyze what's going on. Only from the examples will you be able to catch on.

We have found many old texts of ancient Babylonian mathematics. Among them is a great library full of mathematics workbooks for students. And it's interesting: the Babylonians could solve quadratic equations; they even had tables for solving cubic equations. They could do triangles (See Fig. 2-3); they could do all kinds of things, but they never wrote down

FIGURE 2-3 Pythagorean triples in the Plimpton 322 tablet from about 1700 B.C.

an algebraic formula. The ancient Babylonians had no way of writing formulas; instead, they did one example after the other—that's all. The idea was you're supposed to look at examples until you get the idea. That's because the ancient Babylonians didn't have the power of expression in mathematical form.

Today we do not have the power of expression to tell a student how to understand physics *physically!* We can write the laws, but we still can't say how to understand them physically. The only way you can understand physics physically, because of our lack of machinery for expressing this, is to follow the dull, Babylonian way of doing a whole lot of problems until you get the idea. That's all I can do for you. And the students who *didn't* get the idea in Babylonia flunked, and the guys who *did* get the idea died, so it's all the same!

So, now we try.

2-6 Understanding physics physically

The first problem that I mentioned in Chapter 1 involved a lot of physical things. There were two rods, a roller, a pivot, and a weight—it was 2 kg, I believe. The geometrical relation of the rods was 0.3, 0.4, and 0.5, and

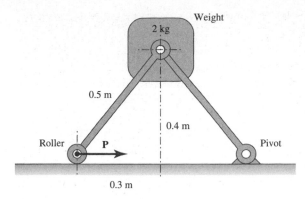

FIGURE 2-4   The simple machine of Chapter 1.

the problem was, what is the horizontal force **P** required at the roller to hold the weight up, as shown in Figure 2-4? It took a little fiddling around (in fact, I had to do it twice before I got it right), but we found that the horizontal force on the roller corresponded to a weight of $\frac{3}{4}$ kg, as shown in Figure 2-5.

Now, if you just let yourself loose of the equations and think about it a while, and you pull back your sleeves and wave your arms, you can almost understand what the answer's going to be—at least *I* can. Now, I have to teach *you* how to do that.

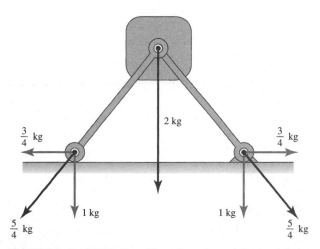

FIGURE 2-5   Distribution of force from the weight, through the rods, to the roller and pivot.

You could say, "Well, the force from the weight comes straight down, and it corresponds to 2 kg, and the weight is balanced equally on two legs. So the vertical force from each leg must be enough to hold up 1 kg. Now, the corresponding horizontal force on each leg must be the fraction of the vertical force that is merely the horizontal to vertical ratio in this right triangle, which is 3 to 4. Therefore, the horizontal force on the roller corresponds to $\frac{3}{4}$ kg weight—period."

Now, let's see if it makes sense: according to that idea, if the roller were shoved much closer to the pivot, so that the distance between the legs was much smaller, I would expect much less force on the roller. Is it true, that when the weight is waaaaay up there, the force on the roller should be low? Yeah! (See Fig. 2-6.)

If you can't *see* it, it's hard to explain *why*—but if you try to hold something up with a ladder, say, and you get the ladder directly *under* the thing, it's easy to keep the ladder from sliding out. But if the ladder is leaning way out at an angle, it's damn hard to keep the thing up! In fact, if you go waaaaay out, so that the far end of the ladder is only a very tiny distance from the ground, you'll find a nearly infinite horizontal force is required to hold the thing up at a very slight angle.

Now, all these things you can *feel*. You don't *have* to feel them; you *can* work them out by making diagrams and calculations, but as problems get more and more difficult, and as you try to understand nature in more and more complicated situations, the more you can guess at, feel, and understand *without actually calculating,* the *much* better off you are! So that's what you should practice doing on the various problems: when you have time somewhere, and you're not worried about getting the answer for a quiz or something, look the problem over and see if you can understand the way it behaves, *roughly,* when you change some of the numbers.

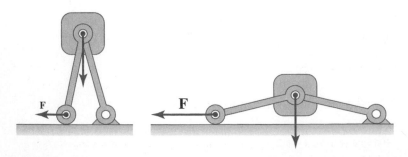

FIGURE 2-6 The force on the roller varies with the height of the weight.

Now, how to explain how to do that, I don't know. I remember once trying to teach somebody who was having a great deal of trouble taking the physics course, even though he did well in mathematics. A good example of a problem that he found impossible to solve was this: "There's a round table on three legs. Where should you lean on it, so the table will be the most unstable?"

The student's solution was, "Probably on top of one of the legs, but let me see: I'll calculate how much force will produce what lift, and so on, at different places."

Then I said, "Never mind calculating. Can you imagine a real table?"

"But that's not the way you're supposed to do it!"

"Never mind how you're *supposed* to do it; you've got a *real* table here with the various legs, you see? Now, where do you think you'd lean? What would happen if you pushed down directly over a leg?"

"Nothin'!"

I say, "That's right; and what happens if you push down near the edge, halfway between two of the legs?"

"It flips over!"

I say, "OK! That's better!"

The point is that the student had not realized that these were not just mathematical problems; they described a real table with legs. Actually, it wasn't a *real* table, because it was perfectly circular, the legs were straight up and down, and so on. But it *nearly* described, *roughly speaking,* a real table, and from knowing what a *real* table does, you can get a very good idea of what *this* table does without having to calculate anything—you know darn well where you have to lean to make the table flip over.

So, how to *explain* that, I don't know! But once you get the idea that the problems are not mathematical problems but *physical* problems, it helps a lot.

Now I'm going to apply this approach to a series of problems: first, in machine design; second, to motions of satellites; third, to the propulsion of rockets; fourth, to beam analyzers, and then, if I still have time, to the disintegration of pi mesons, and a couple of other things. All these problems are pretty difficult, but they illustrate various points as we go along. So, let's see what happens.

2-7 A problem in machine design

First, machine design. Here's the problem: there are two pivoted rods, each a half a meter long, which carry a weight of 2 kg—sound familiar?—and at the left a roller is being driven back or forth by some machinery at a constant velocity of 2 meters per second, OK? And the question for you is, *what is the force required to do that when the height of the weight is 0.4 meters?* (See Fig. 2-7.)

You might be thinking, "We *did* that already! The horizontal force required to balance the weight was $\frac{3}{4}$ of a 1 kg weight."

But I argue, "The force is *not* $\frac{3}{4}$ kg, because the weight is *moving*."

You might counter, "When an object is moving, is a force required to keep it moving? No!"

"But a force *is* required to *change* the object's motion."

"Yes, but the roller is moving at a constant velocity!"

"Ah, yes, that's true: the roller *is* moving at a constant velocity of 2 meters per second. But what about the weight: is *that* moving at a constant velocity? Let's *feel* it: does the weight move slowly sometimes, and fast sometimes?"

"Yes . . ."

"Then its motion is *changing*—and that's the problem we have: to figure out the force required to keep the roller moving constantly at 2 meters per second when the weight is at a height of 0.4 meters."

Let's see if we can understand how the weight's motion is changing.

Well, if the weight is near the top and the roller is almost directly underneath it, the weight hardly moves up and down. In this position the weight is *not* moving very fast. But if the weight is down low, like we had before, and you push the roller just a shade to the right—boy, that weight has to move way up to get out of the way! So, as we push the roller, the weight starts moving up very fast, and then slows down, correct? If it's going up very fast and it gets slower, which way is the acceleration, then? The acceleration must be *down:* it's like I threw it up fast and it slowed down—like it's falling, sort of, so that the force must be *reduced.* That is, the horizontal force I'm going to

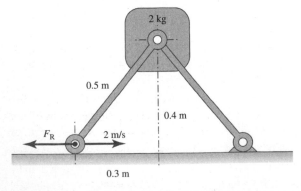

FIGURE 2-7 The simple machine, in motion.

get on the roller is going to be less than it would be if it weren't moving. So we have to figure out how much less. (The reason I went through all this is that I couldn't keep the signs right in the equations, so I had at the end to figure out which way the sign was by this physical argument.)

Incidentally, I have done this problem about four times—making a mistake every time—but I have, at last, got it right. I appreciate that when you do a problem the first time, there are many, many things that get confused: I got the numbers mixed up, I forgot to square, I put the sign of the time wrong, and I did a lot of other things wrong, but anyway, *now* I have it right, and I can show you how it can be done correctly—but I must admit, frankly, that it took me quite a while to get it right. (Boy, I'm glad I've still got my notes!)

Now, in order to calculate the force, we need the acceleration. It's impossible to find the acceleration by just looking at the diagram, with all dimensions fixed at the time of interest. To find the rate of change, we can't leave it fixed—I mean, we can't say, "Well, this is 0.3, this is 0.4, this is 0.5, this is 2 meters per second, what's the acceleration?" There's no easy way to get at that. The only way to find the acceleration is to find the general motion and differentiate it with respect to time.[5] Then we can put in the value of the time that corresponds to this particular diagram.

So I need, therefore, to analyze this thing in a more general circumstance, when the weight is at some arbitrary position. Let's say the pivot and the roller are together at time $t = 0$, and that the distance between them is $2t$, because the roller is moving at 2 meters per second. The time when we want to make the analysis is 0.3 seconds *before* they're together, which is $t = -0.3$, and so the distance between them is actually *negative* $2t$—but it'll be all right if we use $t = 0.3$ and let the distance be $2t$. There will be a lot of signs wrong at the end, but because of my little fishing around at the beginning as to what the right sign was for the force, I'll be all right—I'd rather leave the mathematics alone and get the sign right from physics, than the other way around. Anyhow, here we are. (Don't *you* do this; it's too difficult—it takes practice!)

(Remember what the t means: t is the time before the pivots are together, which is sort of a negative time, which will make everybody crazy, but I can't help it—this is the way I did it.)

Now, the geometry is such that the weight is always (horizontally) halfway between the roller and the pivot. So, if we put the origin of our coordinate system at the pivot, then the x coordinate of the weight is $x = \frac{1}{2}(2t) = t$. The length of the rods is 0.5, so for the height of the weight,

---

[5]See *Alternate Solution A* on page 87 for a way to find the acceleration of the weight without differentiating.

its y coordinate, I got $y = \sqrt{0.25 - t^2}$, by the Pythagorean theorem. (See Fig. 2-8.) Can you imagine, the first time I worked this problem out, very carefully, I got $y = \sqrt{0.25 + t^2}$?

Now we need the acceleration, and the acceleration has two components: one is the horizontal acceleration, and the other is the vertical acceleration. If there's a horizontal acceleration, then there's a horizontal force, and we've got to chase *that* down through the rod and figure out what it is on the roller. This problem is a little easier than it looks because there *is* no horizontal acceleration—the x coordinate of the weight is always half that of the roller; it moves in the same direction, but at half its speed. So, the weight moves horizontally at a constant 1 meter per second. There's no acceleration sideways, thank god! That makes the problem a little easier; we only have to worry about the up and down acceleration.

Therefore to get the acceleration, I must differentiate the height of the weight twice: once to get the velocity in the y direction, and again, to get the acceleration. The height is $y = \sqrt{0.25 - t^2}$. You should be able to differentiate this *fast,* and the answer is

$$y' = \frac{-t}{\sqrt{0.25 - t^2}}. \tag{2.18}$$

It's negative, even though the weight is moving *up.* But I got my signs all bungled up, so I'll leave it *this* way; anyway, I know the speed is up, so this would be wrong if t were positive, but t should really be negative—so it's right anyway.

Now, we calculate the acceleration. There are several ways you can do this: You can do it using ordinary methods, but I'll use the new "super" method I showed you in Chapter 1: you write down y' again; then you say,

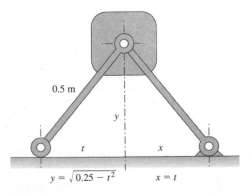

FIGURE 2-8   Using the Pythagorean Theorem to find the height of the weight.

"The first term that I want to differentiate is to the first power, $-t$. Derivative of $-t$ is -1. The next term that I want to differentiate is to the minus one-half power; the term is $0.25 - t^2$. The derivative is $-2t$. *Done!*"

$$y' = -t(0.25 - t^2)^{-1/2}$$

$$y'' = -t(0.25 - t^2)^{-1/2}\left[1 \cdot \frac{-1}{(-t)} - \frac{1}{2} \cdot \frac{-2t}{(0.25 - t^2)}\right] \qquad (2.19)$$

Now we have the acceleration at any time. In order to find the force, we need to multiply it by the mass. So, the force—that is, the extra force besides gravity that's involved because of the acceleration—is the mass, which is 2 kilograms, times this acceleration. Let's put the numbers into this thing: t is 0.3. The square root of $0.25 - t^2$ is the square root of 0.25 minus 0.09, which is 0.16, the square root of which is 0.4—well, how convenient! Is that right? Yes indeed, sir; this square root is the same as y itself, and when t is 0.3, according to our diagram, y is 0.4. OK, no mistake.

(I'm always checking things while I calculate because I make so many mistakes. One way to check it is to do the mathematics very carefully; the other way to check it is to keep seeing whether the numbers that come out are sensible, whether they describe what's really happening.)

Now we calculate. (The first time I did this I put $0.25 - t^2 = 0.4$ instead of 0.16—it took me a while to find *that* one!) We get some number[6] or other, which I have worked out; it's about 3.9.

So, the acceleration is 3.9, and now for the force: the vertical force that this acceleration corresponds to is 3.9 times 2 kilograms times g. No, that's not right! I forgot there's no g now; 3.9 is the true acceleration. The vertical force of gravity is 2 kg times the acceleration due to gravity, 9.8—that's g—and the vertical component of the force of the rod on the weight is the sum of these two, with a minus sign for one; the relative signs are opposite. So, you subtract, and you get

$$F_w = ma - mg = 7.8 - 19.6 = -11.8 \text{ newtons}. \qquad (2.20)$$

But remember, now, this is the *vertical* force on the *weight*. How much is the *horizontal* force on the *roller*? The answer is, the horizontal force on the roller is three-quarters of one-half of the vertical force on the weight. We noticed that before: the force pulling down is balanced by the two legs, which divides it by two, and then the geometry is such that the ratio of the horizontal component to the vertical component is $\frac{3}{4}$—and so the answer is that the horizontal force on the roller is three-eighths of the vertical force

---

[6]The exact number is 3.90625.

on the weight. I worked out the three-eighths of each of these things, and I got 7.35 for gravity, and 2.925 for the term due to the acceleration, and the difference is 4.425 newtons—about 3 newtons less than the force required to support the weight in the same position when it was not moving. (See Fig. 2-9.)

Anyway, that's how you design machines; you know how much force you need to drive that thing forward.

Now, you say, is that the correct way do to it?

There is no such thing! There is no "correct" way to do anything. A particular way of doing it may be correct, but it is not *the* correct way. You can do it any damn way you want! (Well, excuse me: there are *incorrect* ways to do things . . .)

Now, if I were sufficiently smart, I could just look at this thing and tell you what the force is, but I'm *not* sufficiently smart, so I had to do it *some* way or other—but there are *many* ways of doing it. I will illustrate one other way, which is very useful, especially if you are involved in designing real machines. This problem is somewhat simplified by having the legs equal, and so on, because I didn't want to complicate the arithmetic. But the *physical ideas* are such that you can figure the whole thing out another way, even when the geometry is not so simple. And that is the following, interesting, other way.

When you have a whole lot of levers moving a lot of weights, you can do this: as you drive the thing along, and all the weights begin to move

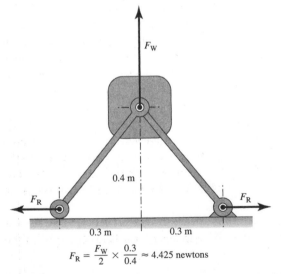

$$F_R = \frac{F_W}{2} \times \frac{0.3}{0.4} \approx 4.425 \text{ newtons}$$

FIGURE 2-9 Using similar triangles to find the force on the roller.

because of all the levers, you're doing a certain amount of work, W. At any given time there's a certain power going in, which is the rate at which you are working, dW/dt. At the same time, the energy of all the weights, E, is changing at some rate, dE/dt, and those should match each other; that is, the rate at which you put work in should match the rate of change of the total energy of all of the weights:

$$\frac{dE}{dt} = \frac{dW}{dt}. \tag{2.21}$$

As you may recall from the lectures, power is equal to force times velocity:[7]

$$\frac{dW}{dt} = \frac{\mathbf{F} \cdot d\mathbf{s}}{dt} = \mathbf{F} \cdot \frac{d\mathbf{s}}{dt} = \mathbf{F} \cdot \mathbf{v}. \tag{2.22}$$

And so, we have

$$\frac{dE}{dt} = \mathbf{F} \cdot \mathbf{v}. \tag{2.23}$$

The idea, then, is that at a given instant the weights have some kind of a speed, and thus they have a kinetic energy. They also have a certain height above the ground, and so they have a potential energy. So if we can figure out how fast the weights are moving and where they are, in order to get their total energy, and then we differentiate that with respect to time, that would be equal to the product of the component of force in the direction that the thing being worked on is moving, times its speed.

Let's see if we can apply that to our problem.

Now, when I push on the roller with a force $\mathbf{P} = -\mathbf{F_R}$ while moving it at a velocity $\mathbf{v_R}$, the rate of change of the energy of the whole darn thing, with respect to time, should equal the *magnitude* of the force times the speed, $F_R v_R$, because in this case the force and the velocity are both in the same direction. It's not a general formula; if I had asked you for the force *in some other direction,* I couldn't have gotten it by this argument directly because this method only gives you the component of the force that does the work! (Of course, you can get it indirectly because you can know the force is going along the rod. If there were several more rods connected, this method would still work, provided you took the force in a direction of motion.)

What about all the work done by all the forces of the constraints—the roller, the pivots, and all the other machinery that holds this stuff in the

---

[7]See *FLP* Vol. I, Chapter 13.

right motion? *No* work is done by them, provided *they* aren't worked on by *other* forces as they go along. For example, if somebody else is sitting over there, pulling one leg out while I'm pushing the other one in, I've got to take the work done by the other guy into account! But nobody's doing that, so, with $v_R = 2$, we have

$$\frac{dE}{dt} = 2F_R. \tag{2.24}$$

So I'm all set if I can calculate dE/dt—divide by two, and lo and behold: the *force!*

Ready? Let's *go!*

Now, we have the total energy of the weight in two pieces: kinetic energy plus potential energy. Well, the potential energy is easy: it's *mgy* (see Table 2-3). We already know that y is 0.4 meters, m is 2 kg, and g is 9.8 meters per second squared. So the potential energy is $2 \times 9.8 \times 0.4 = 7.84$ joules. And now the kinetic energy: well, after a lot of fiddling around, I'll get the velocity of the weight, and I'll write in the kinetic energy for that; we'll do that in just a second. Then I'm all set because I'll have the total energy.

I'm *not* all set: unfortunately, I don't *want* the energy! I need the *derivative* of the energy with respect to time, and you cannot find how fast something changes by figuring out how much it is right *now!* You've either got to figure it out at two adjacent times—now, and an instant later—or, if you want to use the mathematical form, you figure it out for an arbitrary time, t, and differentiate with respect to t. It depends on which is the easiest to do: it may be numerically much easier to figure out the geometry for two positions than it is to figure out the geometry in general, and to differentiate.

(Most people immediately try to put a problem in mathematical form and differentiate it because they don't have enough experience with arithmetic to appreciate the tremendous power and ease of doing calculations with numbers instead of letters. Nevertheless, we'll do it with letters.)

Again, we have to solve this problem, where $x = t$, and $y = \sqrt{0.25 - t^2}$, so that we will be able to calculate the derivative.

Now, we need the potential energy. That we can get very easily: it's *mg* times the height, y, and that comes out to

$$P.E. = mgy = 2 \text{ kg} \times 9.8 \text{ m/s}^2 \times \sqrt{0.25 - t^2} \text{ m}$$

$$= 19.6 \text{ newtons} \times \sqrt{0.25 - t^2} \text{ m} \tag{2.25}$$

$$= 19.6\sqrt{0.25 - t^2} \text{ joules}.$$

But more interesting, and harder to figure out, is the kinetic energy. The kinetic energy is $\frac{1}{2}mv^2$. To figure out the kinetic energy, I need to figure out the velocity squared, and that takes a lot of fooling around: the velocity squared is its x component squared plus its y component squared. I could figure out the y component just like I did before; the x component, I've already pointed out, is 1, and I could have squared those and added them together. But supposing I hadn't already done that, and I wanted to think of still *another* way to get the velocity.

Well, after thinking about it, a good machine designer usually can figure it out from the principles of geometry and the layout of the machinery. For example, since the pivot is stationary, the weight must move around it in a circle. So, in which direction must the velocity of the weight be? It can have *no* velocity *parallel* to the rod, because that would change the length of the rod, right? Therefore, the velocity vector is *perpendicular* to the rod. (See Fig. 2-10.)

You might say to yourself, "Ooh! I have to learn that trick!"

No. That trick is only good for a special kind of problem; it doesn't work most of the time. Very rarely do you happen to need the velocity of something that is rotating around a fixed point; there's no rule that says, "velocities are perpendicular to rods," or anything like that. You have to use common sense as often as possible. It's the general idea of analyzing the machine geometrically that's important here—not any specific rule.

So, now we know the direction of the velocity. The horizontal component of the velocity, we already know, is 1, because it's half the speed of the roller. But look! The velocity is the hypotenuse of a right triangle that is

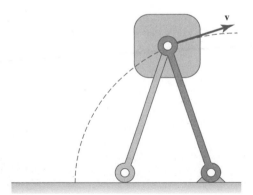

FIGURE 2-10   The weight moves in a circle, so its velocity is perpendicular to the rod.

similar to a triangle having the rod as its hypotenuse! To obtain the magnitude of the velocity is no harder than finding its proportion to its horizontal component, and we can get that proportion from the other triangle, which we already know all about. (See Fig. 2-11.)

Finally, for the kinetic energy we get

$$K.E. = \tfrac{1}{2}mv^2 = \tfrac{1}{2} \times 2\text{kg} \times \left(\frac{0.5}{\sqrt{0.25 - t^2}} \text{ m/s}\right)^2 = \frac{1}{1 - 4t^2} \text{ joules.} \quad (2.26)$$

Now, for the signs: the kinetic energy is certainly positive, and the potential energy is positive because I measured the distance from the floor. So now I'm all right with the signs. So, the energy at any time is

$$E = K.E. + P.E. = \frac{1}{1 - 4t^2} + 19.6\sqrt{0.25 - t^2}. \quad (2.27)$$

Now, in order to find the force using this trick, we need to differentiate the energy and then we can divide by two and everything will be ready. (The apparent ease with which I do this is false: I swear I did it more than once before I got it right!)

Now, we differentiate the energy with respect to time. I'm not going to stall around with this; you're supposed to know how to differentiate by now. So there we are, with the answer for dE/dt (which, incidentally, is twice the force required):

$$\frac{dE}{dt} = \frac{8t}{(1 - 4t^2)^2} - \frac{19.6t}{(0.25 - t^2)^{1/2}}. \quad (2.28)$$

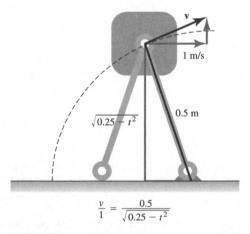

FIGURE 2-11 Using similar triangles to find the velocity of the weight.

So I'm all finished; I need merely put 0.3 in for t, and I'm all done. Well, not quite—to make the signs come out right, I have to use $t = -0.3$:

$$\frac{dE}{dt}(-0.3) = -\frac{2.4}{0.4096} + 19.6 \times \frac{0.3}{0.4}$$

$$\approx 8.84 \text{ watts.}$$

(2.29)

Now, let's see whether this makes sense. If there were no motion, and I didn't have to worry about the kinetic energy, then the total energy of the weight would just be its potential energy, and its derivative should be the force due to the weight.[8] And sure enough, it comes out here the same as we calculated in Chapter 1, 2 times 9.8 times $\frac{3}{4}$.

The first term on the right-hand side of Eq. (2.29) is negative because the weight is decelerating, so it's losing kinetic energy; the second term is positive because the weight is going up, so it's gaining potential energy. Anyhow, they're opposite each other, which is all I want to know, and you can put the numbers in, and sure enough, the force comes out to be the same as before:

$$2F_R = \frac{dE}{dt} \approx 8.84$$

$$F_R \approx 4.42 \text{ newtons.}$$

(2.30)

In fact, this is why I had to do it so many times: after doing it the first time, and being completely satisfied with my wrong answer, I decided to try to do it another, completely different, way. After I did it the other way, I was satisfied with a completely different answer! When you work hard, there are moments when you think, "At last, I've discovered that mathematics is inconsistent!" But pretty soon you discover the error, as I finally did.

Anyway, that's just two ways of solving this problem. There's no unique way of doing any specific problem. By greater and greater ingenuity, you can find ways that require less and less work, but that takes experience.[9]

---

[8]The derivative of the energy with respect to the position of the roller is the magnitude of the force on the roller. However, because the position of the roller happens to equal $2t$ in this particular problem, the derivative of the energy with respect to t equals twice the force on the roller.

[9]See *Alternate Solutions*, beginning on page 87, for three other approaches to solving this problem.

2-8 Earth's escape velocity

I don't have much time left, but the next problem we'll talk about is something involving the motion of planets. I'll have to come back to it because I certainly can't tell you everything about it this time. The first problem is, what is the velocity required to leave the earth's surface? How fast does something have to move so that it can just escape from Earth's gravity?

Now, one way to work that out would be to calculate the motion under the force of gravity, but another way is by the conservation of energy. When the thing reaches way out there, infinitely far away, the kinetic energy will be zero, and the potential energy will be whatever it comes out for infinite distance. The formula for the gravitational potential is in Table 2-3; and it tells us that the potential energy, for particles that are infinitely distant, equals zero.

So, the total energy of something when it leaves Earth at escape velocity must be the same after the thing has gone an infinite distance and Earth's gravity has slowed it down to zero velocity (assuming there are no other forces involved). If M is the mass of the earth, R is the radius of the earth, and G is the universal gravitational constant, we find that the square of the escape velocity must be $2GM/R$.

$$(K.E. + P.E.) \text{ at } \infty, v = 0 \quad = \quad (K.E. + P.E) \text{ at } R, v = v_{\text{escape}}$$

$$(conservation\ of\ energy)$$

$$P.E. \text{ at } \infty \quad = -\frac{GMm}{\infty} = 0 \qquad P.E. \text{ at } R \quad = -\frac{GMm}{R}$$

$$K.E. \text{ at } v = 0 \quad = \frac{m0^2}{2} = 0 \qquad K.E. \text{ at } v = v_{\text{escape}} \quad = \frac{mv_{\text{escape}}^2}{2}$$

$$+ \qquad\qquad\qquad\qquad\qquad +$$

$$0 = \quad \left(-\frac{GMm}{R} + \frac{mv_{\text{escape}}^2}{2} \right)$$

$$\therefore v_{\text{escape}}^2 = \frac{2GM}{R} \tag{2.31}$$

Incidentally, the gravity constant, g (the acceleration of gravity near the earth's surface) is GM/R^2 because the law of force, for a mass, m, is $mg = GMm/R^2$. In terms of the easier-to-remember gravity constant I can write $v^2 = 2gR$. Now, g is 9.8 m/s², and the radius of the earth is 6400 km, so the earth's escape velocity is

$$v_{\text{escape}} = \sqrt{2gR} = \sqrt{2 \times 9.8 \times 6400 \times 1000} = 11{,}200 \text{ m/s.} \tag{2.32}$$

So you have to go 11 kilometers per second to get out—which is pretty fast.

Next, I would talk about what happens if you are going 15 kilometers per second, and you're shooting *past* the earth at some distance.

Now, at 15 kilometers per second, the thing has enough energy to get out, going straight up. But is it obviously necessary that it gets out if it's *not* going straight up? Is it possible that the thing will go around and come back? That's not self-evident; it takes some thought. You say, "It has enough energy to get out," but how do you know? We didn't calculate the escape velocity for *that* direction. Could it be that the sideways acceleration due to Earth's gravity is enough to make it turn around? (See Fig. 2-12.)

It *is* possible, in principle. You know the law that you sweep out equal areas in equal times, so you know that when you get far out, you have to be moving sideways somehow or other. It's not clear that some of the motion that you need to escape isn't going sideways, so that even at 15 kilometers per second you don't escape.

Actually, it turns out that at 15 kilometers per second it does escape—it escapes as long as the velocity is greater than the escape velocity we computed above. As long as it *can* escape, it *does* escape—although that's not self-evident—and the next time, I'm going to try to show it. But to give you a hint as to how I'm going to show it, so you can play around with it yourself, it's the following.

We'll use the conservation of energy at two points, A and B, at its shortest distance from Earth, *a*, and at its longest distance from Earth, *b*, as shown in Figure 2-13; the problem is to calculate *b*. We know the total

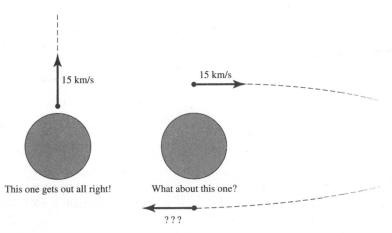

15 km/s

15 km/s

This one gets out all right!   What about this one?

???

FIGURE 2-12   Does having the escape velocity *guarantee* escape?

energy of the thing at A, and it's the same at B because the energy is con-served, so if we knew the velocity at B, we could calculate its potential energy, and thus b. But we don't know the velocity at B!

Yet we do: from the law that equal areas are swept out in equal times, we know that the speed at B must be lower than the speed at A, in a certain proportion—in fact, it's a to b. Using that fact to get the speed at B, we're able to find this distance b in terms of a, and we'll do that next time.

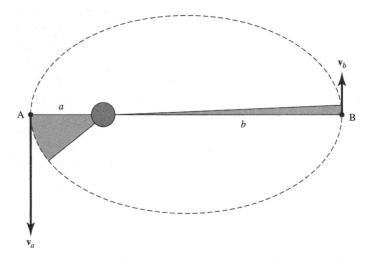

FIGURE 2-13 Satellite distance and velocity at perihelion and aphelion.

Alternate Solutions   *by Michael A. Gottlieb*

Here are three more approaches to solving the machine design problem presented earlier in this chapter (Section 2-7), beginning on p. 73.

A   *Finding the acceleration of the weight using geometry*

The weight is always horizontally halfway between the roller and the pivot, so its horizontal speed is 1 m/s, half the speed of the roller. The weight moves on a circle (centered at the pivot), so its velocity is perpendicular to the rod. By similar triangles we obtain the velocity of the weight. (See Fig. 2-14a.)

Because the weight moves on a circle, the radial component of its acceleration is

$$a_{rad} = \frac{v^2}{r} = \frac{(1.25)^2}{0.5} = 3.125.$$

as per Eq. (2.17). The vertical acceleration of the weight is the sum of its radial and transverse components. (See Fig. 2-14b.)

Using similar triangles again, we obtain the vertical acceleration:

$$a_y = \frac{a_y}{a_{rad}} \times a_{rad} = \frac{0.5}{0.4} \times 3.125 = 3.90625.$$

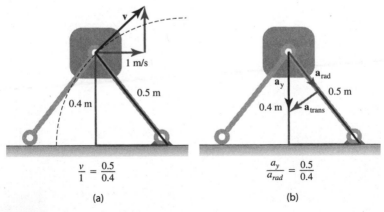

$$\frac{v}{1} = \frac{0.5}{0.4}$$

(a)

$$\frac{a_y}{a_{rad}} = \frac{0.5}{0.4}$$

(b)

FIGURE 2-14

B Finding the acceleration of the weight using trigonometry

The weight moves on a circular arc of radius $\frac{1}{2}$, so its equations of motion can be expressed in terms of the angle the rods make with the ground. (See Fig. 2-15.)

$$x = \tfrac{1}{2}\cos\theta$$
$$y = \tfrac{1}{2}\sin\theta$$

The horizontal speed of the weight is 1 m/s (half the speed of the roller). So $x = t$, $dx/dt = 1$, and $d^2x/dt^2 = 0$. The vertical acceleration can be calculated by differentiating y with respect to t twice. But first, since $t = \frac{1}{2}\cos\theta$,

$$\frac{d\theta}{dt} = -\frac{2}{\sin\theta}.$$

Therefore,

$$\frac{dy}{dt} = \tfrac{1}{2}\cos\theta \cdot \frac{d\theta}{dt} = \tfrac{1}{2}\cos\theta \cdot \left(-\frac{2}{\sin\theta}\right) = -\cot\theta$$

$$\frac{d^2y}{dt^2} = \frac{1}{\sin^2\theta} \cdot \frac{d\theta}{dt} = \frac{1}{\sin^2\theta} \cdot \left(-\frac{2}{\sin\theta}\right) = -\frac{2}{\sin^3\theta}.$$

When $x = t = 0.3$, we have $y = 0.4$ and $\sin\theta = 0.8$ (since $y = \frac{1}{2}\sin\theta$). The magnitude of the vertical acceleration is thus

$$a_y = \left|\frac{d^2y}{dt^2}\right| = \frac{2}{(0.8)^3} = 3.90625.$$

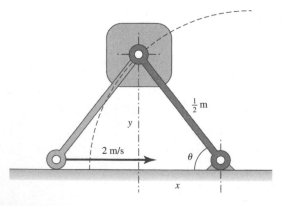

FIGURE 2-15

C Finding the force on the weight using torque and angular momentum

The torque on the weight is $\tau = xF_y - yF_x$. The weight moves horizontally at 1 m/s, so there is no horizontal force on it: $F_x = 0$. Letting $x = t$, the torque reduces to $\tau = tF_y$. Torque is the time derivative of angular momentum, so if we can find the angular momentum L of the weight, we can differentiate it and divide by t to get F_y:

$$F_y = \frac{\tau}{t} = \frac{1}{t}\frac{dL}{dt}.$$

The angular momentum of the weight is easy to find because the weight moves in a circle. Its angular momentum is simply the length of the rod r, times the momentum of the weight, which is its mass m, times its speed v. The speed can be found using Feynman's geometric method (see Fig. 2-16) or by differentiating the weight's equations of motion.

Putting this all together we have:

$$F_y = \frac{1}{t}\frac{dL}{dt} = \frac{1}{t}\frac{d}{dt}(rmv) = \frac{rm}{t} \cdot \frac{d}{dt}\left(\frac{0.5}{\sqrt{0.25 - t^2}}\right)$$

$$= \frac{0.5 \cdot 2}{t} \cdot \frac{0.5t}{(0.25 - t^2)^{3/2}} = \frac{4}{(1 - 4t^2)^{3/2}}$$

At time $t = 0.3$, we have $F_y = 7.8125$. Dividing by 2 kg gives the vertical acceleration we found before: 3.90625.

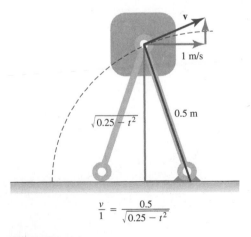

$$\frac{v}{1} = \frac{0.5}{\sqrt{0.25 - t^2}}$$

FIGURE 2-16

3 Problems and Solutions

REVIEW LECTURE C

We're continuing this review of how to do physics by doing a number of problems. All of the problems I chose are elaborate and complicated and difficult; I'll leave you to do the easy problems. Also, I suffer from the disease that all professors suffer from—that is, there never seems to be enough time, and I invented more problems than undoubtedly we'll be able to do, and therefore I've tried to speed things up by writing some things on the board beforehand, with the illusion that every professor has: that if he talks about more things, he'll teach more things. Of course, there's only a finite rate at which material can be absorbed by the human mind, yet we disregard that phenomenon, and in spite of it we go too fast. So, I think I'll just go along slowly, and see how far we get.

3-1 Satellite motion

The last problem that we were talking about was satellite motion. We were discussing the question of whether a particle that was moving perpendicular to the radius of the sun, of a planet, or any other mass M, at a distance a, and having the escape velocity at that distance, would, in fact, escape— because it's not self-evident. It *would* be, if it were headed straight out, radially; but whether it would make it or not if it were headed perpendicular to the radius, is another question. (See Fig. 3-1.)

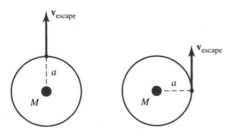

FIGURE 3-1 Escape velocity directed radially and perpendicular to the radius.

It turns out that—if we can remember some of Kepler's laws, and add some other laws like the conservation of energy—we can figure out that if the particle *didn't* escape, it would make an ellipse, and we can figure out how far away it would get, and that's what we're going to do now. If the perihelion of the ellipse is a, how far is the aphelion, b? (By the way, I tried to write this problem on the board, but I found I couldn't spell "perihelion"!) (See Fig. 3-2.)

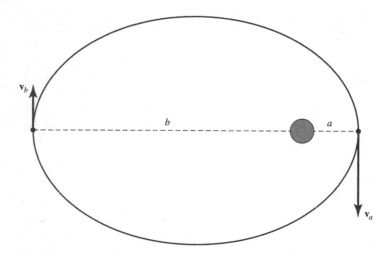

FIGURE 3-2 Velocity and distance at perihelion and aphelion of a satellite in an elliptical orbit.

FIGURE 3-3 Escape velocity from mass M at distance a.

Last time we figured out the escape velocity by using the conservation of energy. (See Fig. 3-3.)

$$\text{K.E.} + \text{P.E. at } a = \text{K.E.} + \text{P.E. at } \infty$$

$$\frac{mv_{\text{escape}}^2}{2} - \frac{GmM}{a} = 0 + 0$$

$$\frac{v_{\text{escape}}^2}{2} = \frac{GM}{a} \tag{3.1}$$

$$v_{\text{escape}} = \sqrt{\frac{2GM}{a}}.$$

Now, this is the formula for the escape velocity at the radius a, but suppose the velocity v_a is arbitrary, and we're trying to find b in terms of v_a. The conservation of energy tells us that the kinetic energy plus the potential energy of the particle at the perihelion must equal the kinetic energy plus the potential energy at the aphelion—and that's what we can use to calculate b, at first sight:

$$\frac{mv_a^2}{2} - \frac{GmM}{a} = \frac{mv_b^2}{2} - \frac{GmM}{b}. \tag{3.2}$$

Infelizmente,[1] however, we do not have v_b, so unless there's some external machinery or analysis to obtain v_b, we're never going to solve Eq. (3.2) for b.

But if we remember Kepler's law of equal areas, we know that in a given time the same area is swept out at the aphelion as is swept out at the perihelion: in a short time Δt the particle at the perihelion moves a distance $v_a\Delta t$ so the area swept out is about $av_a\Delta t/2$, while at the aphelion, where the particle moves $v_b\Delta t$, the area swept out is about $bv_b\Delta t/2$. And so "equal areas" means that $av_a\Delta t/2$ equals $bv_b\Delta t/2$—which means that the velocities vary inversely as the radii. (See Fig. 3-4.)

$$av_a\Delta t/2 = bv_b\Delta t/2$$

$$v_b = \frac{a}{b}v_a. \tag{3.3}$$

That gives us, then, a formula for v_b in terms of v_a, which we can substitute in Eq. (3.2). Then we will have an equation to determine b:

$$\frac{mv_a^2}{2} - \frac{GmM}{a} = \frac{m\left(\frac{a}{b}v_a\right)^2}{2} - \frac{GmM}{b}. \tag{3.4}$$

---

[1] "Unfortunately," in Brazilian Portuguese.

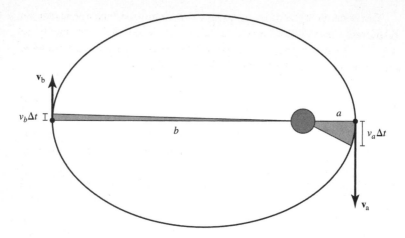

FIGURE 3-4   Using Kepler's law of equal areas to find the velocity of a satellite at aphelion.

Dividing by m, and rearranging, we get

$$\frac{a^2 v_a^2}{2}\left(\frac{1}{b}\right)^2 - GM\left(\frac{1}{b}\right) + \left(\frac{GM}{a} - \frac{v_a^2}{2}\right) = 0. \qquad (3.5)$$

If you look at Eq. (3.5) a while, you could say, "Well, I can multiply by b^2, and then it'll be a quadratic equation in b," or, if you prefer, you could look at it just the way it is, and solve the quadratic equation for $1/b$—either way. The solution for $1/b$ is

$$\begin{aligned}\frac{1}{b} &= \frac{GM}{a^2 v_a^2} \pm \sqrt{\left(\frac{GM}{a^2 v_a^2}\right)^2 + \frac{v_a^2/2 - GM/a}{a^2 v_a^2/2}} \\ &= \frac{GM}{a^2 v_a^2} \pm \left(\frac{GM}{a^2 v_a^2} - \frac{1}{a}\right).\end{aligned} \qquad (3.6)$$

I'm not going to discuss the algebra from here on; you know how to solve a quadratic equation, and there are two solutions for b: one of them is b equals a, it turns out—and that's happy, because if you look at Eq. (3.2) you see it's obvious that if b equals a, the equation will match. (Of course, that doesn't mean that b *is* a.) With the other solution, we get a formula for b in terms of a, which is given here:

$$b = \frac{a}{\dfrac{2GM}{a v_a^2} - 1}. \qquad (3.7)$$

The question is whether we can write the formula in such a way that the relationship of v_a to the escape velocity at the distance a can readily be seen. Notice that by Eq. (3.1) $2GM/a$ is the square of the escape velocity, and therefore we can write the formula this way:

$$ b = \frac{a}{\left(v_{\text{escape}}/v_a\right)^2 - 1}. \tag{3.8} $$

That's the final result, and it is rather interesting. Suppose, first, that v_a is less than the escape velocity. Under those circumstances, we'd expect the particle not to escape, so we should get a sensible value for b. And sure enough, if v_a is less than v_{escape}, then v_{escape}/v_a is greater than 1, and the square is also greater than 1; taking away 1, you get some nice positive number, and a divided by that number tells us b.

To check roughly how accurate our analysis is, a good thing to play around with is the numerical calculation we made of the orbit in the ninth lecture,[2] to see how close the b that we calculated then agrees with the b we get from Eq. (3.8). Why should they not agree perfectly? Because, of course, the numerical method of integration treats time as little blobs instead of continuous, and therefore it isn't perfect.

Anyway, that's how we get b when v_a is less than v_{escape}. (Incidentally, knowing b and knowing a, we know the semi-major axis of the ellipse, and thus we could figure out the period of the orbit from Eq. (3.2), if we wanted to.)

But the interesting thing is this: suppose, first, that v_a is exactly the velocity of escape. Then v_{escape}/v_a is 1, and Eq. (3.8) says that then b is infinite. That means that the orbit is *not* an ellipse; it means that the orbit goes off to infinity. (It can be shown that it is a parabola, in this special case.) So, it turns out, that if you're anywhere near a star or a planet, and no matter what direction you're moving, if you have the velocity of escape, you'll escape, all right—you won't get caught, even though you're not pointed in the right direction.

Still another question is, what happens if v_a *exceeds* the velocity of escape? Then v_{escape}/v_a is less than 1, and b turns out negative—and that doesn't mean anything; there is no real b. Physically, that solution looks more like this: with a very high velocity, much higher than the velocity of escape, a particle coming in is deflected—but its orbit is not an ellipse. It is, in fact, a hyperbola. So the orbits of objects moving around the sun are not only ellipses, as Kepler thought, but the generalization to higher

---

[2]See *FLP* Vol. I, Section 9-7.

speeds includes ellipses, parabolas, and hyperbolas. (We didn't prove here that they are ellipses, parabolas, or hyperbolas, but that's the answer to the problem.)

3-2 Discovery of the atomic nucleus

This hyperbolic orbit business is interesting, and has a very interesting historical application, which I'd like to show you; it is illustrated in Figure 3-5. We take the limiting case of an *enormously* high speed, and a relatively small force. That is, the object is going by so fast that in the first approximation it goes in a straight line. (See Fig. 3-5.)

Suppose we have a nucleus with charge $+Zq_{el}$ (where $-q_{el}$ is the electron charge), and a charged particle that is moving past it at a distance b—an ion of some kind (it was originally done with an alpha particle), it doesn't make any difference; you can put in your own case—let's take a proton of mass m, velocity v, and charge $+q_{el}$ (for an alpha particle, it would be $+2q_{el}$). The proton doesn't go quite in a straight line, but is deflected through a very small angle. The question is, what's the angle? Now, I'm not going to do it exactly, but roughly—to get some idea of how the angle varies with b. (I'll do it nonrelativistically, although it's just as easy to take relativity into account—just a minor change that you can figure out for yourself.) Of course, the bigger b is, the smaller the angle ought to be. And the question is, does the angle decrease as the square of b, the cube of b, as b, or what? We want to get some idea about this.

(This is, as a matter of fact, the way you start on any complicated or unfamiliar problem: you first get a rough idea; then you go back when you understand it better and do it more carefully.)

So the first rough analysis will run something like this: as the proton flies by, there are sideways forces on it from the nucleus—of course, there

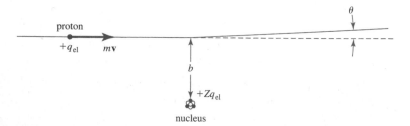

FIGURE 3-5 A high-speed proton is deflected by the electric field as it passes near the nucleus of an atom.

are forces in other directions too, but it's the sideways force that makes it deflect so instead of going straight as it did before, it now has an upward component of velocity. In other words, it acquired some upward momentum as a result of the forces in that direction.

Now, how big is the upward force? Well, it changes as the proton goes along, but more or less, roughly, the force has to depend on b, and the maximum force (as the proton is passing the central position) is

$$\text{vertical force} \approx \frac{Zq_{el}^2}{4\pi\epsilon_0 b^2} = \frac{Ze^2}{b^2}. \tag{3.9}$$

(I substituted e^2 for $\dfrac{q_{el}^2}{4\pi\epsilon_0}$ so I can write the equations quicker.[3])

If I knew how long that force acted, I could estimate the momentum that was delivered. How long *does* the force act? Well, it doesn't act when the proton is a mile away, but, roughly speaking, a force of that general order of magnitude is acting as long as the proton is in the general neighborhood. How far? More or less, when it's passing within a distance b of the nucleus. So the time during which the force acts is of an order of magnitude of the distance b divided by the speed, v. (See Fig. 3-6.)

$$\text{time} \approx \frac{b}{v}. \tag{3.10}$$

Newton's law says that force equals the rate of change of the momentum—so, if we multiply the force by the time over which it's acting,

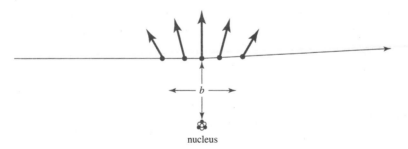

nucleus

FIGURE 3-6 The electric force of the nucleus effectively acts on the proton for a time proportional to the closest distance between them.

---

[3]This historical convention is introduced in *FLP* Vol. I, Section 32-2. Today, the letter e in this context would typically be reserved for the charge on an electron.

we get the change in momentum. Therefore, the vertical momentum acquired by the proton is

$$\text{vertical momentum} = \text{vertical force} \cdot \text{time}$$

$$\approx \frac{Ze^2}{b^2} \cdot \frac{b}{v} = \frac{Ze^2}{bv}. \tag{3.11}$$

That's not *exactly* right; ultimately, when we do an exact integration of this thing, there may be a numerical factor of 2.716 or something—but for now, we're just trying to find the order of magnitude as it depends on the various letters.

The *horizontal* momentum that the particle has when it comes out is, for all intents and purposes, the same as when it went in, which is mv:

$$\text{horizontal momentum} = mv. \tag{3.12}$$

(This is the only thing you need to change to take relativity into account.)

Now, then, what is the angle of deflection? Well, we know the "up" momentum is Ze^2/bv and the "sideways" momentum is mv, and the proportion of "up" to "sideways" is the tangent of the angle—or, practically, the angle itself, since it's so small. (See Fig. 3-7.)

$$\theta \approx \frac{Ze^2}{bv} \bigg/ mv - \frac{Ze^2}{bmv^2}. \tag{3.13}$$

Eq. (3.13) shows how the angle depends on the velocity, on the mass, on the charge, and on the so-called "impact parameter"—the distance b. When you actually calculate θ by integrating the force instead of just estimating it, it turns out that there is indeed a numerical factor missing, and that factor is exactly 2. I don't know whether you've gotten that far in integrations or not: if you can't do it, all right; it's not essential, but the correct angle is

$$\theta = \frac{2Ze^2}{bmv^2}. \tag{3.14}$$

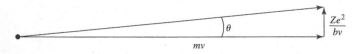

FIGURE 3-7 The horizontal and vertical components of the proton's momentum determine the angle of deflection.

(Actually, you can work the formula out exactly for any hyperbolic orbit, but never mind: you can understand everything for this case, for small angles. Of course Eq. (3.14) is not true when the angles get to 30 or 50 degrees; then we've made too rough an approximation.)

Now, this has a very interesting application in the history of physics—it is the way Rutherford discovered that the atom has a nucleus. He had a very simple idea: by making an arrangement in which alpha particles from a radioactive source would go through a slit—so he knew that they were going in a definite direction—and letting them impinge on a zinc sulfide screen, he could see scintillations in a single spot right behind the slit. But if he put a gold foil between the slit and the screen, the scintillations would sometimes appear elsewhere! (See Fig. 3-8.)

Of course, the reason was, the alpha particles coming past the little nuclei in the gold foil were deflected. By measuring the angles of deflection and using Eq. (3.14) in reverse, Rutherford was able to obtain the distances, b, required to produce that much deflection. The great surprise was, these distances were very much smaller than an atom. Before Rutherford made this experiment it was believed that the positive charge of the atom was not concentrated at a point in the center, but distributed uniformly throughout. Under those circumstances, the alpha particle could never get the big force needed to make the observed deflections, because if it were outside the atom it wouldn't be close enough to the charge, and if it were inside the atom there'd be as much charge above it as below it, and that wouldn't produce enough force. So it was demonstrated by the large deflections that there were sources of strong electric force inside the atom, and then it was guessed that there must be a central point where all the positive charges are, and by observing the deflections as far out as possible, and how many times they occurred, one could

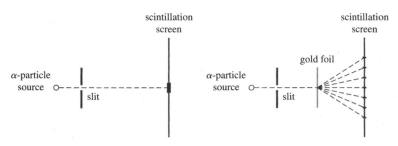

FIGURE 3-8 Rutherford's experiment deflecting alpha particles, which led to the discovery of the atomic nucleus.

obtain an estimate of how small b might be, and ultimately obtain the size of the nucleus—and the size of the nucleus turned out to be 10^{-5} times smaller than the atom! This was the way that it was discovered that nuclei exist.

3-3   The fundamental rocket equation

Now, the next problem I want to talk about is completely different: it has to do with rocket propulsion, and I'm going to take a rocket floating around in empty space first—forgetting all about gravity, and so on. The rocket's built to hold a lot of fuel; it's got some kind of engine by which it squirts fuel out the back—and from the point of view of the rocket, it's always squirting it out at the same speed. It doesn't turn on and off; we start it, and it just keeps squirting stuff out the rear end until it runs out. We'll suppose that the stuff is squirted out at a rate of μ (that's mass per second), and that it goes out at velocity u. (See Fig. 3-9.)

You might say, "Aren't those the same thing? You know the mass per second; isn't that the velocity?"

No. I can dump a certain amount of mass per second by taking a great big lump of stuff and putting it quietly out each time, or I can take the same mass and *throw* it out each time. So, you see, they're two independent ideas.

Now, the question is, how much velocity will the rocket accumulate after a time? Suppose, for instance, that it uses up 90 percent of its weight: that is, when it's finished using all its fuel the mass of the shell that's left is one-tenth as great as the mass of the whole thing loaded before it started. What speed will the rocket acquire?

Anybody in his right mind would say that it is impossible to get any faster than the speed u, but that's not true, as you'll see in a moment. Maybe you'll say that's perfectly obvious; well, all right. But it is, in fact, true for the following reason.

Let's look at the rocket at any moment, moving at any speed at all. If we move along with the rocket and watch for a time Δt, what do we see? Well, there's a certain mass Δm that goes out—which is, of course, the rocket's

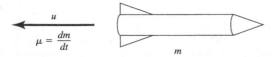

FIGURE 3-9   Rocket with mass m, ejecting fuel at rate $\mu = dm/dt$ with velocity u.

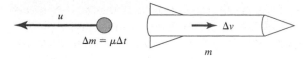

FIGURE 3-10 Rocket gaining speed Δv during interval Δt by ejecting mass Δm with velocity u.

rate of loss μ times the time Δt. And the velocity that this mass comes out at is u. (See Fig. 3-10.)

Now, the moment after this mass is thrown back, how fast is the rocket moving forward? The speed at which it's moving forward must be such that the total momentum is conserved. That is to say, it picks up a little speed, Δv, in such a manner that, if the mass of the rocket shell and remaining fuel at that instant is m, then m times Δv matches the outgoing momentum during that time, which is Δm times u. And that's all there is to the theory of rockets; that's the fundamental rocket equation:

$$m\Delta v = u\Delta m. \tag{3.15}$$

We could put in $\mu\Delta t$ for Δm, and by fiddling around, find out how *long* it takes to get up to a given velocity,[4] but *our* problem is to find the final velocity, and we can do that directly from Eq. (3.15):

$$\frac{\Delta v}{\Delta m} = \frac{u}{m}$$

$$dv = u\frac{dm}{m}. \tag{3.16}$$

In order to find the velocity that the rocket acquires, starting from rest, you integrate $u(dm/m)$ from the initial mass to the final mass. Now, u was assumed constant, so it can be taken outside the integral, and we have, therefore,

$$v = u \int_{m_{\text{initial}}}^{m_{\text{final}}} \frac{dm}{m}. \tag{3.17}$$

---

[4]If the rocket starts at time $t = 0$ with mass $m = m_0$, and $\mu = dm/dt$ is constant, then $m = m_0 - \mu t$, and Eq. (3.16) becomes $dv = u\mu\, dt/(m_0 - \mu t)$. Integrating yields $v = -u \ln\left[1 - (\mu t/m_0)\right]$, and solving for t gives the time required to reach speed v: $t(v) = (m_0/\mu)(1 - e^{-v/u})$.

The integral of *dm/m* may or may not be known to you; let's suppose that it isn't. You say, "1/*m* is such a simple function, I *must* know the derivative: I'll fiddle around with differentiating things until I find it."

But it turns out you can't find anything that's simple—in terms of *m*, powers of *m*, and things like that—which, when you differentiate it, gives 1/*m*. So, not knowing how to do it that way, we'll do it a different way. We'll do it by numerical integration.

Remember: Whenever you're stuck in a mathematical analysis, you can always do it by arithmetic!

3-4　A numerical integration

Let's suppose that the initial mass is 10, and take as a simple approximation that we drop one unit of mass at a time. Furthermore, let's measure all the velocities in terms of the unit *u*, because then we will have simply $\Delta v = \Delta m/m$.

We want to find the total accumulated velocity. Well, let's see: during the first dropping of one unit of mass, how much speed is acquired? Well, that's easy; it's

$$\Delta v = \frac{\Delta m}{m} = \frac{1}{10}.$$

But that isn't exactly right, because while you're spitting one unit of mass out, the mass that's reacting is *not* 10; when you're all finished spitting it out, it's only 9. You see, after Δm is shot out, the mass of the rocket is only $m - \Delta m$, so maybe it would be better to put

$$\Delta v = \frac{\Delta m}{m - \Delta m} = \frac{1}{9}.$$

But that isn't exactly right either. It would be true if the rocket were really throwing out blobs, but it's not—it's dumping mass continuously. At the beginning the mass of the rocket is 10. At the end of the one unit going out, the mass is only 9—so on average, it's more or less like 9.5. During the time the first unit is dropped, we'll say that $m = 9.5$ is the effective average inertia that reacts against the $\Delta m = 1$, so that the rocket receives an impulse Δv equal to 1/9.5:

$$\Delta v \approx \frac{\Delta m}{m - \Delta m/2} = \frac{1}{9.5}.$$

It helps to put these halves in, because then you need fewer steps to get high accuracy. Of course, it still isn't exact. If we wanted to do it more carefully, we could use smaller blobs of mass, like $\Delta m = 1/10$, and do much more analysis. But we'll do it roughly, with $\Delta m = 1$, and keep on going.

Now the mass of the rocket is only 9. We drop another unit off the rear end of the thing, and we find next that Δv is ... 1/9? No.... 1/8? No! It's $\Delta v = 1/8.5$ because the mass has been continuously changing from 9 to 8, and on the average it was roughly 8.5. For the next unit we get $\Delta v = 1/7.5$, and so we discover that the answer is the sum of 1/9.5, 1/8.5, 1/7.5, 1/6.5, ta, ta, ta, ta, tum—to the end. With the last step we go from 2 units of mass down to 1, on the average the mass is 1.5, and we're left with one unit of mass.

Finally, we calculate all these ratios (which takes only a moment to do; these numbers are all honest; it's easy to figure them out) and merely add them together to get the answer, 2.268, which means that the final velocity v is 2.268 times faster than the velocity of the exhaust u. That's the answer to this one—nothin' to it!

1/9.5	0.106
1/8.5	0.118
1/7.5	0.133
1/6.5	0.154
1/5.5	0.182
1/4.5	0.222
1/3.5	0.286
1/2.5	0.400
1/1.5	0.667
	2.268

$$v \approx 2.268\,u \tag{3.18}$$

Now you might say, "I don't like the accuracy here—this is a little sloppy. It's all very well to say, 'in the first step the mass changes from 10 to 9, so it's about 9.5'. But in the last step, it changes from 2 to 1 and you've taken that all on the average of 1.5. Wouldn't it be better to split the last step, dropping half a unit at a time, to get a little bit better accuracy?" (This is a technical point of arithmetic.)

Let's see. While the first half a unit goes out, the mass drops from 2 to 1.5; on average it's 1.75, so I take 1/1.75 times a half unit for my $\Delta m/m$. Then I do the same thing for the second half a unit; the mass drops from 1.5 to 1, averaging 1.25:

$$\Delta v \approx \frac{0.5}{(2 + 1.5)/2} + \frac{0.5}{(1.5 + 1)/2} = \frac{0.5}{1.75} + \frac{0.5}{1.25} = 0.686.$$

So you can make an improvement in the last step—you can improve all the rest of them too, the same way, if you want to go to the trouble—and it comes out 0.686 instead of 0.667, which means that our answer was a little bit low. When you figure it out better it comes out $v \approx 2.287u$. The last digit is really not reliable, but our estimate is pretty close, and the exact answer isn't going to be far from 2.3.

Now, I must tell you, that because the integral $\int_1^x dm/m$ is such a simple function and comes up in so many problems, people have made tables of it and given it a name: it's called the natural logarithm, $\ln(x)$. And if you ever look up $\ln(10)$ in a table of natural logs, you will find it's actually 2.302585:

$$v = u \int_1^{10} \frac{dm}{m} = \ln(10)u = 2.302585\, u \qquad (3.19)$$

You can get that many digits of accuracy by the same technique we used, provided you use a much finer spacing like $\Delta m = 1/1{,}000$ or so, instead of 1—and that's precisely what's been done.

Anyway, we did pretty well in no time at all, without knowing anything, and without looking in tables. So, I keep emphasizing that in emergencies you can always do arithmetic.

3-5 Chemical rockets

Now, this question of rocket propulsion is interesting. You'll notice, first of all, that the speed that is finally acquired is proportional to u, the speed of the exhaust. Therefore all kinds of effort has been put into trying to get the exhaust gases to go out as fast as possible. If you burn hydrogen peroxide with this and that, or oxygen with hydrogen or something, then you get a certain chemical energy generated per gram of fuel. And if you design the nozzles and whatnot correctly, you can get a high percentage of that chemical energy to go into the outgoing velocity. But you can't get more than 100 percent, naturally, and so there's an upper limit for a given fuel as to what speed can be acquired by the most ideal design with a given mass ratio, because there's an upper limit to the value of u that can be acquired from a given chemical reaction.

Consider two reactions, a and b, which have the same energy per atom liberated, but atoms of different masses, m_a and m_b. Then, if u_a and u_b are the exhaust velocities, we have

$$\frac{m_a u_a^2}{2} = \frac{m_b u_b^2}{2}. \qquad (3.20)$$

The velocities will therefore be higher for the reaction with the lighter atom, because whenever $m_a < m_b$, Eq. (3.20) implies that $u_a > u_b$. That's why most of the fuels used in rockets are light materials. The engineers would like to burn helium with hydrogen, but unfortunately that mixture doesn't burn, so for instance, they make do with oxygen and hydrogen.

3-6 Ion propulsion rockets

Instead of using chemical reactions, another proposal is to make a device by which you ionize atoms, and accelerate them electrically. Then you can get a *terrific* velocity, because you can accelerate the ions as much as you want. And so I have another problem here for you.

Suppose we have a so-called ion propulsion rocket. Out of the rear end we are going to squirt cesium ions, accelerated by an electrostatic accelerator. The ions start at the front of the rocket, and a voltage V_0 has been applied between the front and the rear end—in our particular problem, it's not an unreasonable voltage—I took $V_0 = 200,000$ volts.

Now, the problem is, what thrust is this going to produce? It's a different problem than we had before, which was to find how fast would the rocket go. This time, we would like to know what force is produced if the rocket is held in a test stand. (See Fig. 3-11.)

The way it works is this: Suppose that in a time Δt the rocket were to shoot an amount of mass $\Delta m = \mu \Delta t$ at velocity u. Then the momentum going out is $(\mu \Delta t)u$; since action equals reaction, that much momentum is being poured into the rocket. In the other problem the rocket was in space, and so it took off. This time, it's held by the test stand, and the momentum per second that is acquired by the ions is the force that must be applied to hold the rocket in place. The total amount of momentum *per second* acquired by the ions is $(\mu \Delta t)u/\Delta t$. So the thrust force of the rocket is

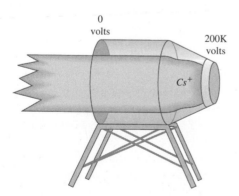

FIGURE 3-11 Ion propulsion rocket on a test stand.

simply μu, the mass per second that is liberated times the velocity at which it goes out. And therefore all I have to do is figure out for my cesium ion what mass per second would go out, and at what velocity:

$$\text{thrust} = \frac{\Delta\,(\text{momentum out})}{\Delta t}$$

$$= (\mu\Delta t)u/\Delta t \tag{3.21}$$

$$= \mu u.$$

We work out the velocity of the ions first, as follows: the kinetic energy of a cesium ion coming out of the rocket is equal to its charge times the voltage difference across the accelerator. That's what voltage is: it's like potential energy, just like field is like force—you just have to multiply by the charge to get the potential energy difference.

The cesium ion is univalent—it has one electron charge—so

$$\frac{m_{Cs^+}u^2}{2} = q_{el}V_0$$

$$u = \sqrt{2V_0\frac{q_{el}}{m_{Cs^+}}}. \tag{3.22}$$

Now, let's figure out this q_{el}/m_{Cs+}. The charge per mole[5] is that famous number 96,500 coulombs per mole. The mass per mole is what's called the atomic weight, and if you look it up in the periodic table, for cesium it's 0.133 kilograms per mole.

You say, "What about these moles? I want to get rid of them!"

They're already gotten rid of: all we need is the *ratio* between the charge and the mass. I can measure that in one atom, or in one mole of atoms, and it's the same ratio. So we get for the outgoing speed

$$u = \sqrt{2V_0\frac{q_{el}}{m_{Cs^+}}} = \sqrt{400{,}000 \cdot \frac{96{,}500}{0.133}} \tag{3.23}$$

$$\approx 5.387 \times 10^5 \text{ m/sec}.$$

Incidentally, 5×10^5 m/s is much faster than you can ever get by a chemical reaction. Chemical reactions correspond to voltages of the order of one volt, and so this ion propulsion rocket provides 200,000 times more energy than a chemical rocket.

Now, that's fine, but we don't want just the velocity; we want the thrust. And so we have to multiply the velocity by the mass per second, μ. I want

---

[5]One mole equals 6.02×10^{23} atoms.

to give the answer in terms of the current of electricity that is pouring out of the rocket—because of course, that's proportional to the mass per second. So, I want to find out how much thrust there is per ampere of current.

Suppose that one ampere is going out: how much mass is that? That's one coulomb per second, or 1/96,500 *moles* per second, because that's how many coulombs are in a mole. But one mole weighs 0.133 kilograms, so it's 0.133/96,500 kilograms per second, and that's the rate of flow of the mass:

$$1 \text{ ampere } = 1 \text{ coulomb/sec} \rightarrow \frac{1}{96,500} \text{ mole/sec}$$

$$\mu = \left(\frac{1}{96,500} \text{ mole/sec}\right) \cdot (0.133 \text{ kg/mole}) \qquad (3.24)$$

$$= 1.378 \times 10^{-6} \text{ kg/sec.}$$

I multiply μ by the speed, u, to find the thrust *per ampere,* and the result is

$$\text{thrust per ampere } = \mu u = (1.378 \times 10^{-6}) \cdot (5.387 \times 10^5)$$
$$\approx 0.74 \text{ newtons/ampere.} \qquad (3.25)$$

So, we get less than three-quarters of a newton per ampere—that's very poor, lousy, low. An ampere isn't a hell of a lot of current, but 100 amperes or 1,000 amperes is quite a job, and it still hardly gives any push. It's hard to get a reasonable amount of ions.

Now let's figure out how much energy is being consumed. When the current is 1 ampere, 1 coulomb of charge per second is dropping through a potential of 200,000 volts. To get the energy (in joules) I multiply the charge by the voltage because volts, really, are nothing but energy per unit charge (joules/coulomb). Therefore $1 \times 200,000$ joules per second is consumed, which is 200,000 watts:

$$1 \text{ coulomb/sec} \times 200,000 \text{ volts} = 200,000 \text{ watts.} \qquad (3.26)$$

We get only 0.74 newtons out of 200,000 watts, which is a pretty punk machine, from an energetic standpoint. The thrust to power ratio is only 3.7×10^{-6} newtons per watt—which is very, very weak:

$$\text{thrust/power } \approx \frac{0.74}{200,000} = 3.7 \times 10^{-6} \text{ newtons/watt.} \qquad (3.27)$$

So, although it's a nice idea, it takes an awful lot of energy to get anywhere in this thing!

3-7 Photon propulsion rockets

Another rocket has been proposed on the basis that the faster you can push the exhaust out the better things are, and so why not push out *photons*— they're the fastest thing on Earth—shoot *light* out the back! You get out there at the rear end of the rocket, you turn on a flashlight, and you get a push! However, you can appreciate that you can pour an awful lot of light out without getting much of a push: you know from experience that when you turn on a flashlight, you don't find yourself thrown off your feet; even if you turn on a 100-watt bulb and put a focuser on it, you don't feel a damn thing! So it's very unlikely that we're going to get much push per watt. Nevertheless, let's try to figure out the thrust-to-power ratio for a photon rocket.

Each photon we throw out the back carries a certain momentum p, and a certain energy E, and the relationship, for photons, is that the energy is the momentum times the speed of light:

$$E = pc. \tag{3.28}$$

So for a photon the momentum per energy is equal to $1/c$. That means that, no matter how many photons we use, the momentum we throw out per second has a definite ratio to the energy we throw out per second—and that ratio is unique and fixed; it's 1 over the speed of light.

But the momentum per second thrown out is the force needed to hold the rocket in place, while the energy per second thrown out is the power of the engine generating the photons. So the thrust-to-power ratio is also $1/c$ (c being 3×10^8), or 3.3×10^{-9} newtons per watt, which is a thousand times worse than the cesium ion accelerator, and a million times worse than a chemical engine! These are some of the points of rocket design.

I am showing you all these rather complicated semi-new things so you can appreciate that you *have* learned *something*, and that you can now understand a great deal of what goes on in the world.

3-8 An electrostatic proton beam deflector

Now, the next problem that I cooked up, to show you how you can do things, is the following. In the Kellogg Laboratory,[6] we have a Van de Graaff generator that generates protons at 2 million volts. The potential difference is generated electrostatically by a moving belt. The protons drop through this potential, pick up a lot of energy, and come out in a beam.

---

[6]The Kellogg Radiation Laboratory at Caltech performs experiments in nuclear physics, particle physics, and astrophysics.

Suppose, for certain experimental reasons, we would like the protons to come out at a different angle, so that we need to deflect them. Now, the most practical way to do this is with a magnet; nevertheless, we can also work out how it can be done electrically—they have been made that way—and that's what we're going to do now.

We take a pair of curved plates that are very close together compared to the radius of their curvature—say they're about $d = 1$ cm apart, separated by insulators. The plates are curved in a circle, and we put as high a voltage as we can across them, from a voltage supply, so that we get an electric field in between that deflects the beam radially, around the circle. (See Fig. 3-12.)

In fact, if you put much more than 20 kilovolts across a 1 cm gap in a vacuum, you have breakdown troubles—whenever there is a little leak, dirt gets in and it's very hard to keep it from sparking over—so let's say we put 20 kilovolts across the plates. (However, I'm not going to *do* this problem with numbers; I'm just *explaining* it all with the numbers, so I'll call the voltage across the plates V_p.) Now, we would like to know: to what radius of curvature do we have to bend the plates so that 2 MeV protons will be deflected between them?

This simply depends on the centripetal force. If m is the mass of a proton, then Eq. (2.17) tells us that mv^2/R equals the force that's needed to pull it in. And the force that we have pulling it in is the charge of the proton—which is again our famous q_{el}—multiplied by the electric field that's in between the plates:

$$q_{el}\mathcal{E} = m\frac{v^2}{R}. \tag{3.29}$$

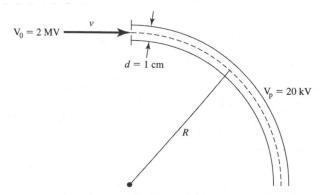

$V_0 = 2$ MV v

$d = 1$ cm

$V_p = 20$ kV

R

FIGURE 3-12   Electrostatic proton beam deflector.

This equation is Newton's law: you have force equals mass times acceleration. In order to use it, however, you've got to know the velocity of the protons coming out of the Van de Graaf generator.

Now, information on the velocity of the protons comes from our knowledge of how much potential they have fallen through—2 million volts—which I'll call V_0. The conservation of energy tells us that the kinetic energy of the proton, $mv^2/2$, equals the charge of the proton multiplied by the voltage through which it has fallen. We can calculate v^2 directly from this:

$$\frac{mv^2}{2} = q_{el}V_0$$

$$v^2 = \frac{2q_{el}V_0}{m}. \tag{3.30}$$

When I substitute v^2 from Eq. (3.30) into Eq. (3.29), I get

$$q_{el}\mathcal{E} = m\frac{\left(\dfrac{2q_{el}V_0}{m}\right)}{R} = \frac{2q_{el}V_0}{R} \tag{3.31}$$

$$R = \frac{2V_0}{\mathcal{E}}.$$

So if I knew what the electric field between the plates was, I could easily find the radius—because of this simple relationship between the electric field, the voltage at which the protons started, and the curvature of the plates.

Well, what *is* the electric field? If the plates don't bend too much, the electric field is approximately the same everywhere between them. And when I put a voltage across the plates, there's an energy difference between a charge on one plate and a charge on the other. The energy difference per unit charge is the voltage difference—that's what voltage *means*. Now, if I carried a charge q from one plate to the other through a constant electric field $\mathcal{E}$, the force on the charge would be $q\mathcal{E}$, and the energy difference would be $q\mathcal{E}d$, where d is the distance between the plates. By multiplying force times distance I get energy—or by multiplying *field* times distance, I get *potential*. So the voltage on the plates is $\mathcal{E}d$:

$$V_p = \frac{\text{energy difference}}{\text{charge}} = \frac{q\mathcal{E}d}{q} = \mathcal{E}d \tag{3.32}$$

$$\mathcal{E} = V_p/d.$$

I have therefore substituted $\mathcal{E}$ from Eq. (3.32) into Eq. (3.31) and by fiddling around, I get the formula for the radius—it's $2V_0/V_p$ times the distance between the plates:

$$R = \frac{2V_0}{(V_p/d)} = 2\frac{V_0}{V_p}d. \tag{3.33}$$

In our particular problem, the ratio of V_0 to V_p—2 million volts to 20 kilovolts—is 100 to 1, and $d = 1$ centimeter. Therefore the radius of curvature should be 200 cm, or 2 meters.

An assumption that's been made here is that the electric field between the plates is constant. If the electric field isn't constant, how good is our deflector? Pretty good anyway, because with a 2-meter radius, the plates are almost flat, so the field *is* nearly constant, and if we've got the beam right in the middle, it's just right. But even if we don't, it's very good because if the field is too strong on one side, it'll be too weak on the other, and those things will compensate, nearly. In other words, by using the field near the middle, we're getting an excellent estimate: even if it's not perfect, it's damn close for such dimensions; at $R/d = 200$ to 1, it's almost exact.

3-9   Determining the mass of the pi meson

I have no more time, but I'll ask you to stay just a minute extra, so I can tell you about one more problem: this is historically the way the mass of the pi meson (π) was determined. In fact, the pi meson was first *discovered* on photographic plates in which there were tracks of mu mesons[7] (μ): some unknown particle had come in and stopped, and where it stopped, there was a little track coming off whose properties were found to be those of a mu meson. (Mu mesons were known before, but the pi meson was just discovered from these pictures.) It was presumed that a neutrino (ν) went off in the opposite direction (leaving no track, because it is neutral). (See Fig. 3-13.)

The rest energy of the μ was known to be 105 MeV, and its kinetic energy was found from the properties of the track to be 4.5 MeV. Supposing all that, how can we find the mass of the π? (See Fig. 3-14.)

Let's suppose that the π is at rest, and that it disintegrates into a μ and a neutrino. We know the rest energy of the μ, as well as the kinetic energy

---

[7]"Mu meson" is an obsolete term for a *muon,* an elementary particle with the same charge as an electron but approximately 207 times the mass (and which in fact isn't a meson at all in the modern meaning of the word "meson").

FIGURE 3-13 Tracks of a pi meson that disintegrated into a muon and an unseen (electrically neutral) particle.

FIGURE 3-14 Disintegration of a pi meson at rest into a muon and a neutrino having equal and opposite momenta. The total energy of the muon and neutrino equals the rest energy of the pi.

of the μ, and therefore the total energy of the μ. But we also need to know the energy of the neutrino because, by relativity, the mass of the π times c squared is its energy, and all that energy goes into the μ and the neutrino. You see, the π disappears, and the μ and the neutrino are left, and by the conservation of energy, the energy of the π must be the energy of the μ plus the energy of the neutrino:

$$E_\pi = E_\mu + E_\nu. \tag{3.34}$$

So we need to calculate both the energy of the μ and the energy of the neutrino. The energy of the μ is easy; it's practically given: it's 4.5 MeV kinetic, added to the rest energy—so you get $E_\mu = 109.5$ MeV.

Now what's the energy of the neutrino? That's the hard one. But by the conservation of momentum, we know the *momentum* of the neutrino because it's exactly equal and opposite to the momentum of the μ—and that's the key. You see, I'm running it backwards here: if we knew the momentum of the neutrino, we could probably figure out its energy. So, let's try.

We calculate the momentum of the μ from the formula $E^2 = m^2c^4 + p^2c^2$, choosing a system of units for which $c = 1$, so that $E^2 = m^2 + p^2$. Then, for the momentum of the μ we get

$$p_\mu = \sqrt{E_\mu^2 - m_\mu^2} = \sqrt{(109.5)^2 - (105)^2} \approx 31 \text{ MeV}. \tag{3.35}$$

But the momentum of the neutrino is equal and opposite, so—not worrying about signs, only magnitude—the momentum of the neutrino is also 31 MeV.

What about its energy?

Because the neutrino has zero rest mass, its energy equals its momentum times c. We talked about that for the "photon rocket." For this problem we let $c = 1$, so the energy of the neutrino is the same as its momentum, 31 MeV.

Well, we're all finished: the energy of the μ is 109.5 MeV, the energy of the neutrino is 31 MeV, so the total energy liberated in the reaction was 140.5 MeV—all given by the rest mass of the π:

$$m_\pi = E_\mu + E_\nu \approx 109.5 + 31 = 140.5 \, \text{MeV}. \qquad (3.36)$$

And this is the way that the mass of the π was originally determined.

That's all I have time for. Thank you.

See you next term. Best of luck!

4 *Dynamical Effects and Their Applications*

I just want to announce that the lecture I give today is unlike the others, in that I'll talk about a large number of subjects which are only for your own entertainment and interest, and if you don't understand something because it's too complicated, you can just forget about it; it's absolutely unimportant.

Every subject that we study could, of course, be studied in greater and greater detail—certainly in greater detail than would be warranted for a first approach—and we could continue to pursue the problems of rotational dynamics almost forever, but then we wouldn't have time to learn much else about physics. So we're going to take leave of the subject here.

Now, someday you may want to return to rotational dynamics, each in your own way, whether as a mechanical engineer, or an astronomer worrying about the spinning stars, or in quantum mechanics (you have rotation in quantum mechanics)—however it comes back to you again, that's up to you. But this is the first time that we will leave a subject unfinished; we have a lot of broken ideas, or threads of ideas, that go out and aren't continued, and I'd like to tell you where they go, so that you get some better appreciation of what you know.

In particular, most of the lectures up until now have been, to a large extent, theoretical—full of equations, and so on—and many of you with an interest in practical engineering may be longing to see a few instances of the "cleverness of man" in making use of some of these effects. If that's so, our subject today is ideally suited to delight you, because there's nothing more exquisite in mechanical engineering than the practical development of inertial guidance over the last few years.

This was dramatically illustrated by the voyage of the submarine *Nautilus* under the polar ice cap: no stars could be observed; maps of the bottom of the sea, under the ice cap, were practically non-existent; inside the ship there was no way to see where you were—and nevertheless they knew at any moment exactly where they were.[1] The trip would have been impossible without the development of inertial guidance, and I would like to explain to you today how it works. But before I get to that, it will be better if I explain a few of the older, less sensitive devices in order for you to more fully appreciate the principles and problems involved in the more delicate and marvelous developments that came later.

4-1 A demonstration gyroscope

In case you haven't seen one of these things, Figure 4-1 shows a demonstration gyroscope, set in gimbals.

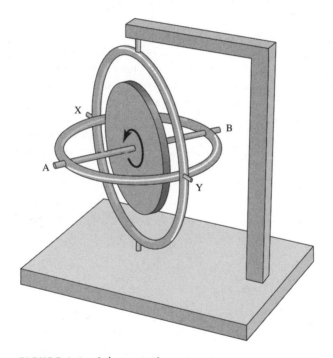

FIGURE 4-1 A demonstration gyroscope.

---

[1] In 1958, USS *Nautilus,* the world's first nuclear-powered submarine, sailed from Hawaii to England, passing the North Pole on August 3. It was under the polar ice cap for a total of 95 hours.

Once the wheel is set spinning it stays in the same orientation even if the base is picked up and moved around in an arbitrary direction—the gyroscope remains with its spin axis, AB, fixed in space. For practical applications, where the gyro must be kept spinning, a small motor is used to compensate for friction in the gyro's pivots.

If you try to change the direction of axis AB by pushing downward on point A (creating a torque on the gyro around axis XY), point A does not move downward but actually moves sideways, towards Y in Figure 4-1. Applying a torque to the gyro around any axis (other than the spin axis) produces a rotation of the gyro around an axis that is mutually perpendicular to the applied torque and the gyro's spin axis.

4-2   The directional gyro

I start with the simplest possible application of a gyroscope: if it's in an airplane which is turning from one direction to another, the gyro's axis of rotation—set horizontally, for example—stays pointed in the same direction. This is very useful: as the airplane goes through various motions, you can maintain a direction—it's called a directional gyro. (See Fig. 4-2.)

You say, "That is like a compass."

It is not like a compass, because it doesn't seek north. It's used like this: when the airplane is on the ground, you calibrate the magnetic compass and

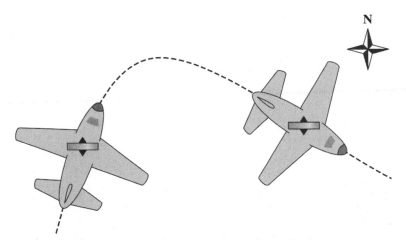

FIGURE 4-2   A directional gyroscope maintains its orientation in an airplane that is turning.

use it to set the gyro's axis in some direction, say north. Then as you fly around, the gyro maintains its orientation, and so you can always use it to find north.

"Why not just use the magnetic compass?"

It's very difficult to use a magnetic compass in an airplane because the needle swings and dips from the motion, and there's iron and other sources of magnetic fields in the airplane.

On the other hand, when the airplane quiets down and goes in a straight line for a while, you'll find that the gyro doesn't point north anymore, because of friction in the gimbals. The airplane has been turning, slowly, and there has been friction, small torques have been generated, the gyro has had precessional motions, and it is no longer pointing in exactly the same direction. So, from time to time it's necessary for the pilot to reset his directional gyro against the compass—every hour, or perhaps more, depending on how perfectly frictionless the thing is made.

4-3   The artificial horizon

The same system works with the artificial horizon, a device to determine "up." When you're on the ground, you set a gyro with its axis vertical. Then you go up in the air, and the airplane pitches and rolls; the gyro maintains its vertical orientation, but it also needs to be reset every once in a while.

What can we check the artificial horizon against?

We could use gravity to find out which way is up, but as you can well appreciate, when you're going in a curve, the apparent gravity is off at an angle, and it's not so easy to check. But in the long run, on average, the gravity *is* in a certain direction—unless the airplane ultimately ends up flying upside down! (See Fig. 4-3.)

And so, consider what would happen if we added a weight to the gimbals at point A of the gyro shown in Figure 4-1, then set the gyro spinning with its axis vertical, and A down. When the plane flies straight and level, the weight pulls straight down which tends to keep the spin axis vertical. As the airplane goes around a corner, the weight tries to pull the axis off vertical, but the gyroscope resists through the precession and the axis drifts away from vertical only very slowly. Eventually the airplane stops its maneuver, and the weight pulls straight down again. In the long run, on average, the weight tends to orient the axis of the gyro in the direction of gravity. This is much like the comparison of the directional gyro to the magnetic compass, except instead of being done every hour or so, it's done perpetually, all during the flight, so that in spite of the gyro's tendency to drift very slowly, its

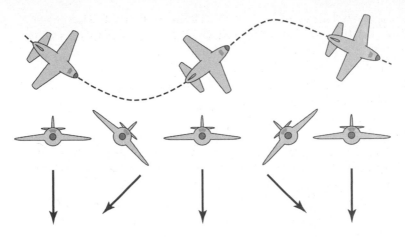

FIGURE 4-3 Apparent gravity in an airplane that is turning.

orientation is maintained by the *average* effect of gravity over long periods of time. The slower the gyro drifts, naturally, the longer the period of time over which this average is effectively taken, and the better the instrument is for more complex maneuvers. It's not unusual to make maneuvers in an airplane that throw gravity off for half a minute, so if the averaging period were only half a minute, the artificial horizon wouldn't work right.

The devices I have just described—the artificial horizon and the directional gyro—are the machinery used to guide automatic pilots in airplanes. That is, information taken from these devices is used to steer the airplane in a certain direction. If, for example, the airplane turns away from the axis of the directional gyro, electrical contacts are made which, working their way through a lot of things, result in some flaps being moved, steering the airplane back on course. Automatic pilots have at their heart such gyroscopes.

4-4   A ship-stabilizing gyroscope

Another interesting application of gyroscopes that is no longer used today, but was once proposed and built, is to stabilize ships. Of course everybody thinks you do this just by spinning a big wheel on an axle affixed to the ship, but that's not right. If you were to do that with the spin axis vertical, for example, and a force pitched the front of the ship up, the net result would be to make the gyro precess to one side, and the ship would flip over—so that doesn't work! A gyroscope doesn't stabilize anything by itself.

FIGURE 4-4 A ship-stabilizing gyroscope: pitching the gyroscope forward creates a torque that rolls the ship to the right.

What's done instead illustrates a principle used in inertial guidance. The trick is this: somewhere in the ship there is a very small, but beautifully built, *master* gyroscope, with its axis, say, vertical. The moment the ship rolls a little bit out of vertical, electrical contacts in the master gyro operate a tremendous *slave* gyro that is used to stabilize the ship—these were probably the biggest gyroscopes ever built! (See Fig. 4-4.) Ordinarily the slave gyro's axis is kept vertical, but it is gimbaled so it can be swiveled around the pitch axis of the ship. If the ship starts to roll right or left, then to straighten it out, the slave gyro is wrenched *back* or *forward*—you know how gyros are always obstinate and go the wrong way. The sudden rotation around the pitch axis produces a torque about the roll axis that opposes the roll of the ship. The *pitching* of the ship is not corrected by this gyro, but of course the pitching of a big ship is relatively small.

4-5 **The gyrocompass**

I'd like now to describe another device used on ships, the "gyrocompass." Unlike the directional gyro, which always drifts away from north and must be reset periodically, a gyrocompass actually seeks north—in fact,

View from above the North Pole:

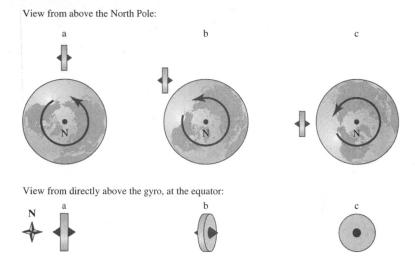

View from directly above the gyro, at the equator:

FIGURE 4-5 A free gyroscope rotating with the earth maintains its orientation in space.

it's better than the magnetic compass because it seeks *true* north, in the sense of the axis of the earth's rotation. It works as follows: suppose that we look at the earth from above the North Pole, going around counter-clockwise, and we have set up a gyroscope somewhere, say on the equator, with its axis pointing east-west, parallel to the equator, as shown in Figure 4-5(a). For the moment let's just take the example of an ideal free gyroscope, with lots of gimbals and whatnot. (It could be in a ball floating in oil—however you want it so that there is no friction.) Six hours later, the gyroscope would be still pointing in the same absolute direction (because there are no torques on it from friction), but if we were standing next to it on the equator, we would see it slowly turning over: six hours later it would be pointing straight up, as shown in Figure 4-5(c).

But now imagine what would happen if we put a weight on the gyro-scope as shown in Figure 4-6; the weight would tend to keep the spin axis of the gyroscope perpendicular to gravity.

As the earth rotates, the weight will be lifted, and the weight lifting up, of course, will want to come back down, and that will produce a torque par-allel to the earth's rotation which will make the gyroscope turn at right angles to everything; in this particular case, if you figure it out, it means that instead of lifting the weight up, the gyro will turn over. And so it turns its axis around toward the north, as shown in Figure 4-7.

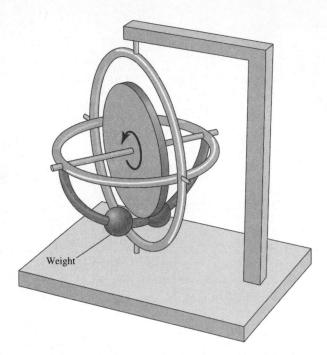

FIGURE 4-6 The demonstration gyroscope with weights that tend to keep the spin axis perpendicular to gravity.

View from above the North Pole:

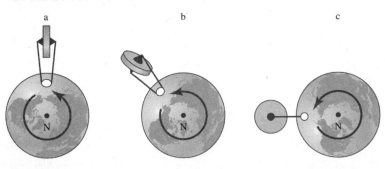

View from directly above the gyro, at the equator:

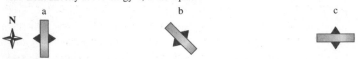

FIGURE 4-7 A weighted gyrocompass tends to align its spin axis parallel to Earth's spin axis.

View from above the North Pole:

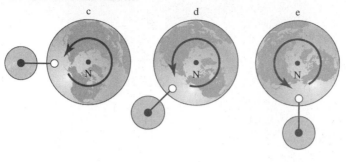

View from directly above the gyro, at the equator:

FIGURE 4-8   A gyrocompass with its spin axis parallel to Earth's tends to stay that way.

Now, suppose the gyro's axis is finally pointing north: will it stay there? If we draw the same picture with the axis pointing north, as shown in Figure 4-8, then as the earth rotates, the arm swings around the gyro's axis and the weight stays down; there are no torques on the axis from the weight being lifted, and the axis is still pointing north later.

So, if the gyrocompass has its axis pointing north, there's no reason why it can't stay that way, but if its axis is pointing even slightly east-west, then as the earth rotates, the weight will turn the axis toward the north. This, therefore, is a north-seeking device. (Actually, if I built it *just* this way, it would seek north and pass, coast on the other side, and go back and forth— so a little bit of damping has to be introduced.)

We have made an artificial gyrocompass kind of a gadget, which is shown in Figure 4-9. The gyroscope unfortunately hasn't got *all* the axes free; it's got two of them free, and you have to do a little thinking to figure out that that's almost the same. You turn the thing around to simulate the motion of the earth, and gravity is imitated by a rubber band tied to the gyro, analogous to the weight on the end of the arm. When you start turning the thing around, the gyro precesses for a while, but if you're patient enough, and keep the thing going, it settles down. The only place where it can stay without trying to turn in some other direction is parallel to the axis of rotation of its frame—the imaginary earth in this case—and so it settles down, very nicely, pointing toward the north. When I stop the rotation, the

FIGURE 4-9   Feynman demonstrates an artificial gyrocompass.

axis drifts, because there are various frictions and forces in the bearings. Real gyros always drift; they don't do the ideal thing.

4-6   Improvements in gyroscope design and construction

The best gyros that could be made about ten years ago had a drift of between 2 and 3 degrees in an hour—that was the limitation of inertial guidance: it was impossible to determine your direction in space more accurately than that. For instance, if you went on a trip in a submarine for 10 hours, the axis of your directional gyro could be off by as much as 30 degrees! (The gyrocompass and the artificial horizon would work all right, because they are "checked" by gravity, but the free-rotating directional gyros wouldn't be accurate.)

The development of inertial guidance required the development of much better gyroscopes—gyroscopes in which the uncontrollable frictional forces that tend to make them precess are at an absolute minimum. A number of inventions have been made to make this possible, and I'd like to illustrate the general principles involved.

In the first place, the gyros we've been talking about so far are "two-degrees-of-freedom" gyroscopes, because there are two ways that the spin axis can turn. It turns out to be better if you only need to worry about one way at a time—that is, it is better to set up your gyros so that you only need to consider the rotations of each one about a single axis. A "one-degree-of-freedom" gyroscope is illustrated in Figure 4-10. (I have to thank Mr. Skull of the Jet Propulsion Laboratory for not only lending me these slides, but also explaining to me everything that's been going on the last few years.)

The gyro wheel is spinning around a horizontal axis ("Spin axis" in the figure), which is only allowed to turn freely around one axis (IA), not two. Nevertheless, this is a useful device for the following reason: imagine that the gyro is being turned around the vertical input axis (IA), because it's in a car or a ship which is turning. Then the gyro wheel will try to precess around the horizontal output axis (OA); more accurately, a torque will be developed about the output axis, and if the torque is not opposed, the gyro wheel will precess about that axis. So, if we have a signal generator (SG) which can detect the angle through which the wheel precesses, then we can use it to discover that the ship is turning.

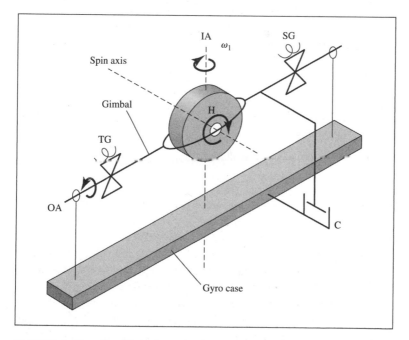

FIGURE 4-10 Simplified schematic of a one-degree-of-freedom gyroscope. Based on original lecture slide.

Now, there are several features to be taken into account here: the delicate part is that the torque around the output axis must represent the result of rotation around the input axis with absolute accuracy. Any *other* torques about the output axis are noise, and we have to get rid of them to avoid confusion. And the difficulty is that the gyro wheel itself has some weight, which has to be supported against the weight of the pivots on the output axis—and *those* are the real problem, because they produce a friction which is uncertain and indefinite.

So the first and main trick that improved the gyroscope was to put the gyro wheel in a can and float the can in oil. The can is a cylinder completely surrounded by oil, and free to turn about its axis ("Output axis" in Figure 4-11). The weight of the can, with the wheel and air in it, is exactly the same as the oil it displaces (or as near as it can be made) so that the can is neutrally balanced. That way there's very little weight to support at the pivots, so very fine jewel bearings can be used, like the ones inside a watch, consisting of a pin and a jewel. Jewel bearings can take very little sideways force, but they don't have to take much sideways force in this case—and they have very little friction. So that was the first great improvement: to float the gyro wheel, and use jewel bearings at the pivots that support it.

The next important improvement was to never actually *use* the gyroscope to create any forces—or very great forces. The way we've been talk-

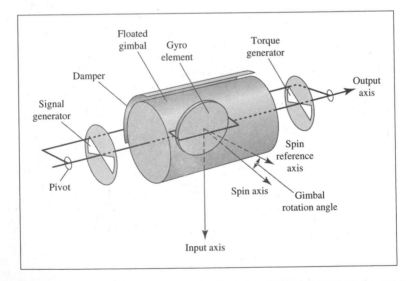

FIGURE 4-11 Detailed schematic of a one-degree-of-freedom integrating gyroscope. Based on original lecture slide.

ing about this thing so far, the gyro wheel precesses about the output axis and we measure how far it precesses. But another interesting technique for measuring the effect of rotation about the input axis is based on the following idea (see Figures 4-10 and 4-11): suppose we have a device carefully built, so that by giving it a definite amount of electric current we can, very accurately, generate a certain torque on the output axis—an electromagnetic torque generator. Then we can make a feedback device with *tremendous* amplification between the signal generator and the torque generator, so that when the ship turns around the input axis, the gyro wheel starts to precess around the output axis, but as soon as it moves a *shade,* a *hair*—just a hair—the signal generator says, "Hey! It's moving!" and the torque generator immediately puts a torque on the output axis that counteracts the torque making the gyro wheel precess, and holds it in place. And then we ask the question, "How hard do we have to hold it?" In other words, we measure the amount of juice going into the torque generator. Essentially, we measure the torque making the gyro wheel precess, by measuring how much torque is needed to counterbalance it. This feedback principle is very important in the design and development of gyroscopes.

Now, another interesting method of feeding back, which is in fact used even more often, is illustrated in Figure 4-12.

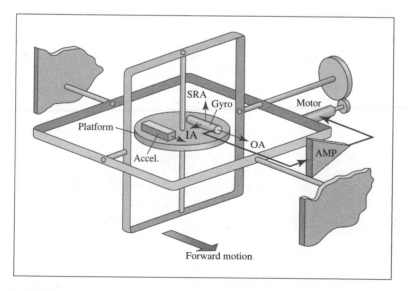

FIGURE 4-12 Schematic of a one-degree-of-freedom stable platform. Based on original lecture slide.

The gyro is the little can ("Gyro" in Figure 4-12) on the horizontal platform (Platform) in the center of the supporting framework. (You can ignore the accelerometer (Accel) for the moment; we'll just worry about the gyro.) Unlike the previous example, this gyro's spin axis (SRA) is vertical; however, the output axis (OA) is still horizontal. If we imagine that the framework is mounted in an airplane traveling in the indicated direction ("Forward motion" in Figure 4-12), then the input axis is the airplane's pitch axis. When the airplane pitches up or down, the gyro wheel starts to precess around the output axis and the signal generator makes a signal, but instead of balancing it by a torque, this feedback system works as follows: as soon as the airplane starts to turn around the pitch axis, the framework which supports the gyroscope in relation to the airplane is turned the opposite way, so as to *undo* the motion; we turn it back, so that we get no signal. In other words, we keep the platform stable via feedback, and we never really move the gyroscope! That's a heck of a lot better than having it swinging and turning, and trying to figure out the airplane's pitch by measuring the output of the signal generator! It's much easier to feed the signal back like this, so that the platform doesn't turn at all, and the gyroscope

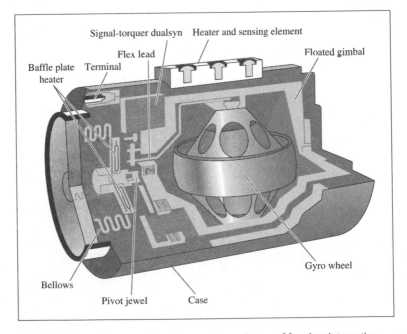

FIGURE 4-13 Cutaway view of an actual one-degree-of-freedom integrating gyroscope. Based on original lecture slide.

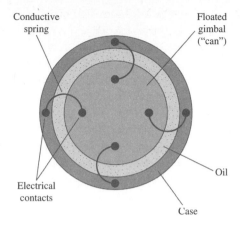

Conductive
spring

Floated
gimbal
("can")

Electrical
contacts

Oil

Case

FIGURE 4-14 Electrical connections from the case to the floated gimbal in a
one-degree-of-freedom gyroscope.

maintains its axis—then we can just *see* the pitch angle, by comparing the
platform to the floor of the airplane.

Figure 4-13 is a cutaway drawing that shows how an actual "one-degree-
of-freedom" gyroscope is built. The gyro wheel looks very big in this picture, but the entire apparatus fits in the palm of my hand. The gyro wheel is
inside of a can, which is floating in a very small amount of oil—it's all in a
little crevice around the can—but it's enough so that no weight needs to be
supported by the minuscule jewel bearings at each end. The gyro wheel is
spinning all the time. The bearings it spins on need not be frictionless,
because they are opposed—the friction is opposed by the engine, which
turns a little motor, which turns the gyro wheel around. There are electromagnetic coils ("Signal-torquer dualsyn" in Figure 4-13) which detect the
very slight motions of the can, and those provide the feedback signals
which are used either to produce a torque on the can around the output axis,
or to turn the platform that the gyro's standing on around the input axis.

There is a technical problem here of some difficulty: to power the motor
that makes the gyro wheel go around, we have to get electricity from a fixed
part of the apparatus into the turning can. That means wires have to come
in contact with the can, yet the contacts must be practically frictionless,
which is very difficult. The way it's done is as follows: four carefully made
springs in semicircular form are connected to conductors on the can, as
shown in Figure 4-14; the springs are made of very good material, like
watch spring material, only very fine. They are balanced so that when the
can is exactly in the zero position they make no torque; if the can is even

slightly rotated, they make a little torque—however, because the springs are so perfectly made, that torque is exactly known—we know the right equations for it—and it's corrected for in the electrical circuits of the feedback devices.

There's also plenty of friction on the can from the oil, which creates torque around the output axis when the can rotates. But the law of friction for liquid oil is very accurately known: the torque is exactly proportional to the speed of the can's rotation. And so it can be completely corrected for in the calculational parts of the circuit that make the feedback, same as the springs.

The big principle of all the accurate devices of this kind is not so much to make everything perfect, but to make everything very definite and precise.

This device is like the wonderful "one-horse shay":[2] everything is made at the absolute limit of mechanical possibilities at the present time, and they're still trying to make it better. But the most serious problem is this: what happens if the gyro wheel's axle is a little off-center in the can, as shown in Figure 4-15? Then the can's center of gravity won't coincide with the output axis, and the weight of the wheel will turn the can around, creating plenty of unwanted torque.

To fix that, the first thing you do is drill little holes, or put weights on the can, to make it as balanced as possible. Then you measure very carefully what remaining drift there is, and use that measurement for calibration. When you've measured a particular device you've built, and find that you can't reduce the drift to zero, you can always correct *that* in the feed-

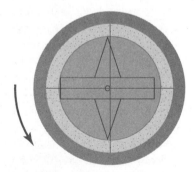

FIGURE 4-15 An unbalanced floated gimbal makes unwanted torque about the output axis in a one-degree-of-freedom gyroscope.

---

[2] *The Deacon's Masterpiece or The Wonderful "One-Hoss Shay": A Logical Story* is Oliver Wendell Holmes' poem about a chaise that was designed so perfectly that it lasted a hundred years and then fell into dust all at once.

back circuit. The problem in this case, though, is that the drift is indefinite: after the gyro runs for two or three hours, the position of the center of gravity moves slightly because of wear in the axle's bearings.

Nowadays, gyroscopes of this kind are over a hundred times better than the ones made 10 years ago. The very best ones have a drift of not more than $1/100^{th}$ of a degree per hour. For the device shown in Figure 4-13, that means the gyro wheel's center of gravity cannot move more than $1/10^{th}$ of one-millionth of an inch from the center of the can! Good mechanical practice is something like 100 millionths of an inch, so this has to be a thousand times better than *good* mechanical practice. Indeed, this is one of the most serious problems—to keep the axle bearings from wearing, so that the gyro wheel moves no more than 20 atoms to either side of the center.

4-7 **Accelerometers**

The devices we've been talking about can be used to tell which way is up, or to keep something from turning around an axis. If we have three such devices set on three axes, with all kinds of gimbals, and so on, then we can keep something absolutely stationary. While the airplane goes around, the platform inside stays horizontal, it never turns to the right or the left; it doesn't do anything. That way we can maintain our north, or east, or up and down, or any other direction. But the next problem is to find out *where we are:* how far have we gone?

Now, you know you can't make a measurement inside an airplane to find out how *fast* it's going, so you certainly can't measure how *far* it's gone, but you *can* measure how much it's *accelerating*. So, if we initially measure no acceleration, we say, "Well, we have zero position and no acceleration." When we start going we have to accelerate. When we accelerate we can measure *that*. And then, if we integrate the acceleration with a calculating machine, we can calculate the speed of the airplane, and, integrating again, we can find its position. Therefore, the method of determining how far something has gone is to measure the acceleration and integrate it twice.

How do you measure the acceleration? An obvious device for measuring acceleration is shown schematically in Figure 4-16. The most important component is just a weight ("Seismic mass" in the figure). There's also a kind of weak spring (Elastic restraint) to hold the weight more or less in place, and a damper to keep it from oscillating, but these details are unimportant. Now, suppose this whole device is accelerated forward, in the direction indicated by the arrow (Sensitive axis). Then, of course, the weight starts to move back, and we use the scale (Scale of indicated accelerations) to measure how far back it moves; from this we can find

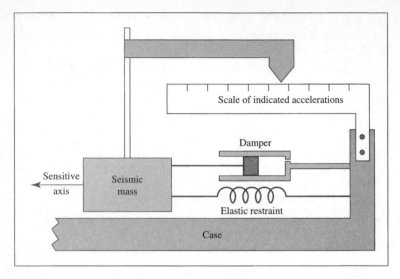

FIGURE 4-16 Schematic of a simple accelerometer. Based on original lecture slide.

the acceleration, and by integrating it twice we get the distance. Naturally, if we make a little error in measuring the position of the weight, so that the acceleration we find is slightly off at some point, then after a long time, over which we integrate twice, the distance is going to be *way* off. So, we have to make the device better.

The next stage of improvement, shown schematically in Figure 4-17, uses our familiar feedback principle: when this device accelerates, the mass moves, and the motion causes a signal generator to output a voltage proportional to the displacement. Then, instead of just measuring the voltage, the trick is to feed it back through an amplifier to a device that pulls the weight back, to find out how much force is needed to keep the weight from moving. In other words, rather than letting the weight move and measuring how far it goes, we measure the reaction force needed to balance it, and then, by $\mathbf{F} = m\mathbf{a}$, we find the acceleration.

One embodiment of this device is shown schematically in Figure 4-18. Figure 4-19 is a cutaway drawing that shows how the actual device is built. It's much like the gyro in Figures 4-11 and 4-13, except the can looks empty: instead of a gyroscope, there's just a weight attached to one side, near the bottom. The whole can is floating so that it is entirely supported and balanced by liquid oil (it's on perfectly beautiful, fine, jewel pivots), and, of course, the weighted side of the can stays down, due to gravity.

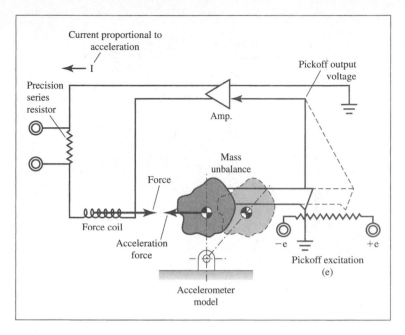

FIGURE 4-17   Schematic of an unbalanced mass accelerometer with force feedback. Based on original lecture slide.

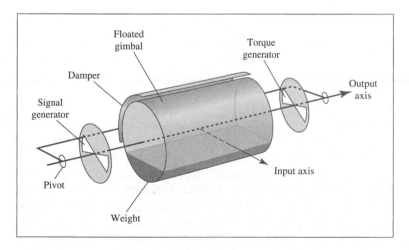

FIGURE 4-18   Schematic of a floated gimbal accelerometer with torque feedback. Based on original lecture slide.

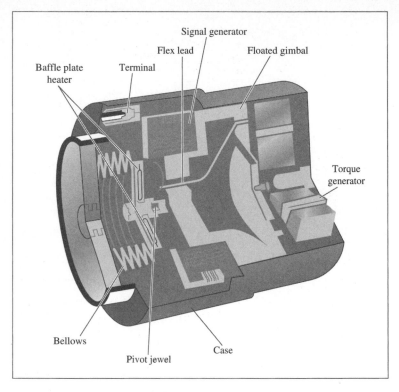

FIGURE 4-19 Cutaway view of an actual floated gimbal accelerometer. Based on original lecture slide.

This device is used to measure horizontal acceleration in the direction perpendicular to the axis of the can; as soon as it accelerates in that direction, the weight lags behind and slops up the side of the can, which turns on its pivots; the signal generator immediately makes a signal, and that signal is put on the torque generator's coils to pull the can back to its original position. Just as before, we feed torque back to straighten things out, and we measure how much is needed to keep the thing from shaking, and that torque tells us how much we're accelerating.

Another interesting device for measuring acceleration, which, in fact, *automatically* does one of the integrations, is shown schematically in Figure 4-20. The scheme is the same as the device shown in Figure 4-11, except that there's a weight ("Pendulous mass" in Figure 4-20) on one side of the spin axis. If this device is accelerated upward, a torque is generated on the

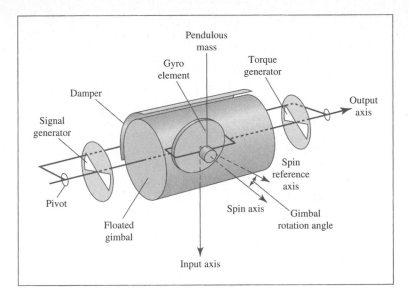

FIGURE 4-20 Schematic of a one-degree-of-freedom pendulous integrating gyro, used as an accelerometer; the gimbal rotation angle indicates velocity. Based on original lecture slide.

gyroscope, and then it's the same as our other device—only the torque is caused by an acceleration, instead of by turning the can. The signal generator, the torque generator, and all the rest of the stuff are the same. The feedback is used to twist the can back around the output axis. In order to balance the can, the upward force on the weight must be proportional to the acceleration, but the upward force on the weight is proportional to the angular velocity at which the can is twisted, so the can's angular velocity is proportional to the acceleration. This implies that the can's *angle* is proportional to *velocity*. Measuring how far the can has turned gives you the velocity—and so one integration is already done. (That doesn't mean this accelerometer is better than the other one; what works best in a particular application depends on a whole lot of technical details, and that's a problem of design.)

4-8   A complete navigational system

Now, if we build some devices like these, we can put them together on a platform as shown in Figure 4-21, which represents a complete navigational system. The three little cylinders (G_x, G_y, G_z) are gyroscopes with

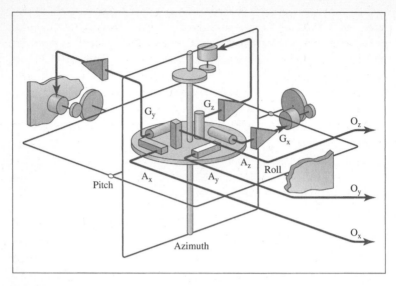

FIGURE 4-21 A complete navigational system, with three gyroscopes and three accelerometers, mounted on a stable platform. Based on original lecture slide.

axes set in three mutually perpendicular directions, and the three rectangular boxes (A_x, A_y, A_z) are accelerometers, one for each axis. These gyroscopes, with their feedback systems, maintain the platform in absolute space without turning in any direction—neither yaw, nor pitch, nor roll—while the airplane (or ship, or whatever it's in) goes around, so that the plane of the platform is very accurately fixed. This is very important for the accelerating-measuring gadgets because you've got to know precisely which directions they're measuring in: if they've gotten cockeyed, so the navigational system thinks they're turned one way when they're actually turned some other way, then the system will go haywire. The trick is to keep the accelerometers in a fixed orientation in space so it's easy to make the displacement calculations.

The outputs of the x, y, and z accelerometers go into integrating circuits, which make the displacement calculations by integrating twice in each direction. So, assuming that we started at rest from a known position, we can know at any moment where we are. And we also know in what direction we're headed, because the platform is still in the same direction it was set when we started (ideally). That's the general idea. However, there are a few points I'd like to make.

First, when measuring acceleration, consider what happens if the device makes an error of, say, one part in a million. Suppose it's in a rocket, and it needs to measure accelerations up to 10 g. It would be hard to resolve less than 10^{-5} g with a device that can measure up to 10 g (in fact, I doubt you could). But it turns out that a 10^{-5} g error in acceleration, after you integrate it twice for an hour, means an error in position of over half a kilometer—after 10 hours, it's more like 50 kilometers, which is *way* off. So this system won't just keep on working. In rockets it doesn't matter because all the acceleration happens at the very beginning and afterwards they coast free. However, in an airplane or a boat you need to reset the system from time to time, just like an ordinary directional gyro, to make sure it is still pointed the same way. This can be done by looking at a star or the sun, but how do you check it inside a submarine?

Well, if we have a map of the ocean, we can see if we went over a mountain top or something that was supposed to pass underneath us. But suppose we don't have a map—there's still a way to check! Here's the idea: the earth is round, and, if we have determined that we've gone, say, 100 miles in some direction, then the gravitational force should no longer be in the same direction as it was before. If we don't keep the platform perpendicular to gravity, the output of the acceleration-measuring devices will be all wrong. Therefore we do the following: we start with the platform horizontal, and use the accelerating-measuring devices to calculate our position; according to the position we figure out how we *should* turn the platform so that it remains horizontal, and we turn it at a rate predicted to keep it horizontal. That's a very handy thing—but it's *also* the device which saves the day!

Consider what would happen if there was an error. Suppose the machine was just standing in a room, not moving, and after some time, because it was built imperfectly, the platform was *not* horizontal, but rotated slightly, as shown on Figure 4-22(a). Then the weights in the

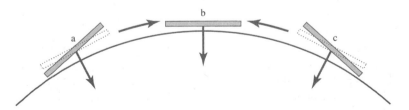

FIGURE 4-22   Earth's gravity is used to check that the stable platform remains horizontal.

accelerometers would be displaced, corresponding to an acceleration, and the positions calculated by the machinery would indicate motion to the right, towards (b). The mechanism which tries to keep the platform horizontal would rotate it slowly, and eventually, when the platform was level again, the machine would no longer think it's accelerating. However, because of the apparent acceleration, the machine would still think it had a velocity in the same direction, and so the mechanism which tries to keep the platform horizontal would continue to rotate it, very slowly, until it was no longer horizontal, as shown in Figure 4-22(c). In fact, it would go through the zero of acceleration, and then it would think it was accelerating in the opposite direction. So we'd have an oscillatory motion which is very small, and the errors would only accumulate over one of these oscillations. If you figure out all the angles and turnings and whatnot, it takes 84 minutes for one of these oscillations. Thus, it is only necessary to make the device good enough to give the right accuracy within a period of 84 minutes, because it will correct itself in that time. It is much like what is done in an airplane where the gyrocompass is checked against a magnetic compass from time to time, but in this case the machine is checked against gravity as in the case of the artificial horizon.

In roughly the same manner, the azimuth device on a submarine (which tells you which way is north) is set from time to time against a gyrocompass, which is averaging over long periods, so that the motions of the ship don't make any difference. Thus, you can correct the azimuth against the gyrocompass, and you can correct the accelerometers against gravity, and so the errors do not accumulate forever, but only for about an hour and a half.

In the *Nautilus* submarine there were three monstrous platforms of this type, each in a great big ball, hung right next to each other from the ceiling of the navigator's room, all completely independent, in case any of them broke down — or, if they didn't agree with each other, the navigator would take the best two out of three (which must have made him pretty nervous!). These platforms were all different when they were built, because you can't make anything perfect. The drift caused by slight inaccuracies had to be measured in each device, and the devices had to be calibrated to compensate for it.

There's a laboratory at JPL where some of these new devices are tested. It's an interesting laboratory, if you consider how you would check such a device: you don't want to get in a ship and move around; no, in this laboratory they check the device against the rotation of the earth! If the device is sensitive, it will turn because of the rotation of the earth, and it will drift. By measuring the drift, corrections can be determined within a very short time. This laboratory is probably the only one in the world whose fundamental feature—the thing that makes it go—is the fact that the earth is turning. It wouldn't be useful for calibration if the earth didn't turn!

4-9 Effects of the earth's rotation

The next thing I want to talk about is effects of the rotation of the earth (besides the effects on the calibration of inertial guidance devices).

One of the most obvious effects of the rotation of the earth is on the large-scale motion of the winds. There's a famous tale, which you hear again and again, that if you have a bathtub, and you pull out the plug, the water goes around one way if you're in the Northern Hemisphere, and the other way if you're in the Southern Hemisphere—but if you try it, it doesn't work. The reason it's *supposed* to go around one way is something like this: suppose we have a plug in a drain at the bottom of the ocean, under the North Pole. Then we pull the plug out, and the water starts moving down the drain. (See Fig. 4-23.)

The ocean has a large radius, and the water is slowly turning around the drain because of the earth's rotation. As the water comes in toward the drain it goes from a larger radius to a smaller radius, and so it has to go around faster to maintain its angular momentum (like when the spinning ice skater pulls her arms in). The water goes around the same way the earth is turning, but it has to turn faster, so somebody standing on the earth would see the water swirling around the drain. That's right, and that's the way it should work. And that's the way it *does* work with the winds: if there's a place where there's low pressure, and the surrounding air is trying to move into it, then instead of moving straight in, it gets some sideways motion—

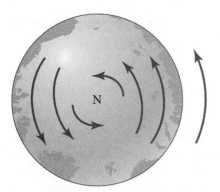

FIGURE 4-23   Water going down an imaginary drain at the North Pole.

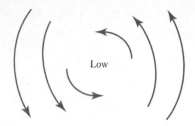

FIGURE 4-24 High pressure air converging into a low pressure zone in the Northern Hemisphere.

in fact, ultimately, the sideways motion becomes so great, that instead of moving in at all, the air is practically rotating around the low pressure area.

So this is one of the laws of weather: if you face downwind in the Northern Hemisphere, low pressure is always on the left, high pressure on the right (see Figure 4-24), and the reason has to do with the rotation of the earth. (This is *nearly* always true; from time to time, under certain crazy circumstances, it doesn't work, because there are other forces involved besides the rotation of the earth.)

The reason it doesn't work in your bathtub is as follows: what causes this phenomenon is the initial rotation of the water—and the water in your bathtub *is* rotating. But how fast is the earth's rotation? Once around a *day.* Can you guarantee that the water in your bathtub hasn't got a little bit of motion equivalent to one swash around the bathtub in a *day?* No. Ordinarily, there's a lot of swishing and swashing in the tub! So this only works on a big enough scale, like a great big lake, where the water's pretty quiet, and you can easily demonstrate that the circulation is not so great as to correspond to once around the lake in a day. Then, if you make a hole in the bottom of the lake and let the water run out, it'll turn in the correct direction, as advertised.

There are a few other points about the rotation of the earth which are interesting. One of them is that the earth is not exactly a sphere; it's a little bit off as a result of its spinning—the centrifugal forces, balancing against gravity, make it oblate. And you can calculate *how* oblate, if you know how much the earth gives. If you assume it's like a perfect fluid that oozes into its ultimate position and ask what the oblateness should be, you'll find that it agrees with the actual oblateness of the earth within the accuracy of the calculations and the measurements (an accuracy of about 1 percent).

This is not true of the moon. The moon is more lopsided than it ought to be, for the speed at which it's turning. In other words, either the moon was turning faster when it was liquefied, and it froze strong enough to resist the

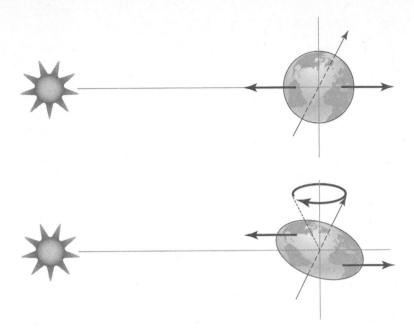

FIGURE 4-25   The oblate earth precesses due to torques induced by gravity.

tendency to get into the right shape, or else it was never molten, but was formed by throwing together a bunch of meteors—and the god who did it didn't do it in a perfectly precise and balanced manner, so it's a little lopsided.

I also want to talk about the fact that the oblate earth is spinning around an axis which is not perpendicular to the plane of the earth's rotation around the sun (or the moon's rotation around the earth, which is almost the same plane). If the earth were a sphere, the gravitational and centrifugal forces on it would be balanced with respect to its center, but because it's a little lopsided, the force is not balanced; there's a torque due to gravitation which tends to turn the earth's axis perpendicular to the line of force, and so, like a great gyroscope, the earth precesses in space. (See Fig. 4-25.)

The axis of the earth, which today points to the North Star, is actually drifting slowly around, and in time it will point to all the stars in the heavens on a big cone subtending an angle of $23\frac{1}{2}$ degrees. It takes 26,000 years for it to come back to the pole star, so if you are reincarnated 26,000 years from now, you may have nothing new to learn, but if it's any other time, you'll have to learn another position (and maybe another name) for the "pole" star.

4-10   The spinning disk

In the last lecture (see *FLP* Vol. I, Ch. 20, "Rotation in Space") we discussed the interesting fact that the angular momentum of a rigid body is not necessarily in the same direction as its angular velocity. We took as an example a disk that is fastened onto a rotating shaft in a lopsided fashion, as shown in Figure 4-26. I'd like to explore this example in further detail.

First, let me remind you of an interesting thing that we've already talked about: that for any rigid body, there is an axis through the body's center of mass about which the moment of inertia is maximal, there is another axis through the body's center of mass about which the moment of inertia is minimal, and these are always at right angles. It's easy enough to see this for a rectangular block as shown in Figure 4-27, but surprisingly it's true for any rigid body.

These two axes, and the axis which is perpendicular to them both, are called the principal axes of the body. The principal axes of a body have

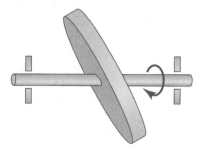

FIGURE 4-26   A disk fastened in a lopsided manner to a spinning shaft.

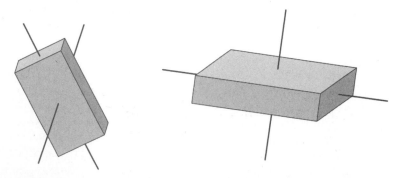

FIGURE 4-27   Rectangular blocks and their principal axes of minimum and maximum moment of inertia.

the following special property: the component of the body's angular momentum in the direction of a principal axis is equal to the component of its angular velocity in that direction times the body's moment of inertia about that axis. So, if $\mathbf{i}$, $\mathbf{j}$, and $\mathbf{k}$ are unit vectors along the principal axes of a body, with respective principal moments of inertia A, B, and C, then when the body rotates about its center of mass with angular velocity $\boldsymbol{\omega} = (\omega_i, \omega_j, \omega_k)$, its angular momentum is

$$\mathbf{L} = A\omega_i\mathbf{i} + B\omega_j\mathbf{j} + C\omega_k\mathbf{k}. \tag{4.1}$$

For a thin disk of mass m and radius r, the principal axes are as follows: the main axis is perpendicular to the disk, with maximal moment of inertia $A = \frac{1}{2}mr^2$; any axis perpendicular to the main axis has the minimum moment of inertia $B = C = \frac{1}{4}mr^2$. The principal moments of inertia are not equal; in fact, $A = 2B = 2C$. So, when the shaft in Figure 4-26 is rotated, the disk's angular momentum is not parallel to its angular velocity. The disk is *statically* balanced because it is attached to the shaft at its center of mass. But it is not *dynamically* balanced. When we turn the shaft, we have to turn the disk's angular momentum, so we must exert a torque. Figure 4-28 shows the disk's angular velocity $\boldsymbol{\omega}$ and its angular momentum $\mathbf{L}$, and their components along the principal axes of the disk.

But now, consider this interesting, additional thing: suppose we put a bearing on the disk, so that we can also spin the disk around *its* main axis with angular velocity $\boldsymbol{\Omega}$, as shown in Figure 4-29.

Then while the shaft is turning, the disk would have an *actual* angular momentum which is the result of the shaft turning *and* the disk spinning. If we spin the disk in the direction opposite to the way the shaft is turning it, as shown in the figure, we will reduce the component of the disk's angular velocity along its main axis. In fact, since the ratio of the disk's principal moments of inertia is exactly 2:1, Eq. 4.1 tells us that by spinning the disk

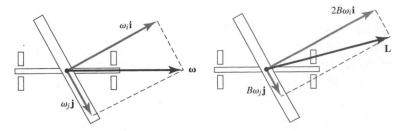

FIGURE 4-28 The angular velocity ω and the angular momentum $\mathbf{L}$ of the disk spun by the shaft, and their components along the principal axes of the disk.

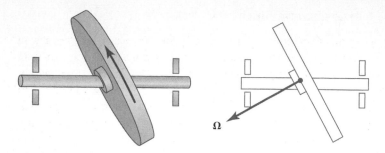

FIGURE 4-29 Spinning the disk around its main axis with angular velocity Ω, while holding the shaft still.

backwards at exactly *half* the speed the shaft turns it around (such that $\Omega = -(\omega_i/2)\,\mathbf{i}$), we can put this thing together in such a miraculous manner that the total angular momentum is exactly along the shaft—and then we can take the shaft away, because there are no forces! (See Fig. 4-30.)

And that is the way a free body turns: if you throw an object into space alone, like a plate[3] or a coin, you see it doesn't just turn around one axis. What it does is a combination of spinning around its main axis, and spinning around some other cockeyed axis in such a nice balance, that the net result is that the angular momentum is constant. That makes it wobble—and the earth wobbles, too.

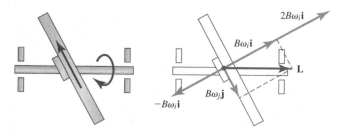

FIGURE 4-30 Spinning the shaft and simultaneously spinning the disk around its main axis in the opposite direction so that the total angular momentum is parallel to the shaft.

---

[3]The spinning/wobbling disk had a special significance to Dr. Feynman, as he writes in "The Dignified Professor" in *Surely You're Joking, Mr. Feynman!*: "The diagrams and the whole business that I got the Nobel Prize for came from that piddling around with the wobbling plate."

4-11   Earth's nutation

From the period of the earth's precession—26,000 years—it's been shown that the maximum moment of inertia (around the pole) and the minimum moment of inertia (around an axis in the equator) differ by only 1 part in 306—the earth is almost a sphere. However, since the two moments of inertia do differ, any disturbance of the earth could result in a slight rotation around some other axis, or, what amounts to the same thing: the earth nutates as well as precesses.

You can calculate the nutation frequency of the earth: it turns out, in fact, to be 306 days. And you can measure it very accurately: the pole wobbles in space by 50 feet measured at the earth's surface; it wobbles around, and back and forth, rather irregularly, but the major motion has a period of 439 days, *not* 306 days, and therein lies a mystery. However, this mystery is easily resolved: the analysis was made for rigid bodies, but the earth is not rigid; it's got liquid goop on the inside, and so, first of all, its period is different from that of a rigid body, and secondly, the motion is damped out so it should stop eventually—that's why it's so small. What makes it nutate at all, despite the damping, are various irregular effects which jiggle the earth, such as the sudden motions of winds, and ocean currents.

4-12   Angular momentum in astronomy

One of the most striking characteristics of the solar system, discovered by Kepler, is that everything goes around in ellipses. This was explained, ultimately, by the law of gravitation. But there are a whole lot of other things about the solar system—peculiar simplifications—which are harder to explain. For example, all the planets seem to go around the sun in roughly the same plane, and, except for one or two, they all rotate around their poles the same way—west to east, like Earth; almost all the planetary moons go around in the same direction, and so with few exceptions, everything turns the same way. It's an interesting question to ask: How did the solar system get that way?

In studying the origin of the solar system, one of the most important considerations is that of angular momentum. If you imagine a whole lot of dust or gas contracting as a result of gravitation, even if it only has a small amount of internal motion, the angular momentum must remain constant; those "arms" are coming in and the moment of inertia is going down, so the angular velocity has to increase. It's possible that the planets are merely the result of a necessity the solar system has to dump its angular momentum from time to time in order to be able to contract still further—we don't

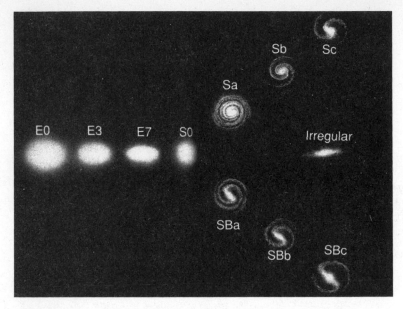

FIGURE 4-31 Different types of nebulae: spiral, barred spiral, and elliptic.

know. But it is true that 95% of the angular momentum in the solar system is in the planets, and not in the sun. (The sun is spinning, all right, but it's only got 5% of the total angular momentum.) This problem has been discussed many times, but it is still not understood how a gas contracts or how a pile of dust falls together when it is rotating slightly. Most discussions pay lip service to the angular momentum at the beginning; then, when they make the analysis, they disregard it.

Another serious problem in astronomy has to do with the development of the galaxies—the nebulae. What determines their form? Figure 4-31 shows several different types of nebulae: the famous ordinary spiral (much like our own galaxy), barred spiral nebulae (whose long arms extend from a central bar), and elliptic nebulae (which don't even have arms). And the question is: How did they become different?

It could be, of course, that the masses of different nebulae are different, and that if you start with different amounts of mass, you come out with different results. That's possible, but because the spiral character of nebulae almost certainly has something to do with angular momentum, it seems more likely that differences from one nebula to another are explained by differences in the initial angular momentum of the original masses of gas and dust (or whatever you assume they start with). Another possibility, which some people have proposed, is that the different types of nebulae represent different stages of development. That would mean that they are all different ages—which, of course, would have dramatic implications for our theory of the universe: Did it all explode at one time, after which the gas condensed to form different types of nebulae? Then they would all have to be the same age. Or, are the nebulae perpetually being formed from debris in space, in which case they could have different ages?

A real understanding of the formation of these nebulae is a problem in mechanics, one involving angular momentum, and one which is still not solved. The physicists should be ashamed of themselves: astronomers keep asking, "Why don't you figure out for us what will happen if you have a big mass of junk pulled together by gravity and spinning? Can you understand the shapes of these nebulae?" And nobody ever answers them.

4-13 Angular momentum in quantum mechanics

In quantum mechanics the fundamental law $\mathbf{F} = m\mathbf{a}$ fails. Nevertheless, some things remain: the law of conservation of energy remains; the law of conservation of momentum remains; and the law of conservation of *angular* momentum also remains—it remains in a very beautiful form, very deep in the heart of the quantum mechanics. Angular momentum is a central feature in the analyses of quantum mechanics, and that's in fact one of the main reasons for going so far into it in mechanics—in order to be able to understand the phenomena in atoms.

One of the interesting differences between classical and quantum mechanics is this: in classical mechanics, a given object can have *arbitrary* amounts of angular momentum by spinning at different speeds; in quantum mechanics, the angular momentum along a given axis cannot be arbitrary— it can only have a value that is an integral or half-integral multiple of Planck's constant over two pi $(h/2\pi, \text{ or } \hbar)$, and it must jump from one value to another in increments of $\hbar$. This is one of the deeper principles of quantum mechanics associated with angular momentum.

Finally, an interesting point: we think of the electron as a fundamental particle, as simple as it can be. Nevertheless, it has an intrinsic angular momentum. We picture the electron not simply as a point charge, but as a point charge that is a sort of limit of a real object that has angular momentum. It is something like an object spinning on its axis in the classical theory, but not exactly: it turns out that the electron is analogous to the simplest kind of gyro, which we imagine to have a very small moment of inertia, spinning extremely fast about its main axis. And, interestingly, the thing that we always do in the first approximation in classical mechanics, which is to neglect the moments of inertia around the precession axis—that seems to be *exactly* right for the electron! In other words, the electron seems to be like a gyroscope with an infinitesimal moment of inertia, spinning at infinite angular velocity, so as to have a *finite* angular momentum. It's a limiting case; it's not *exactly* the same as a gyro—it's even simpler. But it's still a curiosity.

I have here the insides of the gyro shown in Figure 4-13, if you want to look at it. That's all for today.

4-14 **After the lecture**

Feynman: If you look through the magnifier very carefully, you can see the veeeerrrrry fine, semicircular wires which feed power into the can, and are connected to these little pins here, on the outside.

Student: How much does one of these things cost?

Feynman: Oh, god knows how much they cost. There's so much precision work involved, not so much to *make* the thing, but to get it all calibrated and measured. See the tiny holes, and the four gold pins that look like somebody bent them? They bent those pins just so, so that the can would be perfectly balanced. However, if the oil density changes the can won't float: it'll sink in the oil, or rise in the oil, and there'll be forces on the pivots. To keep the oil density right, so the can just floats, you have to keep its temperature right within a few *thousandths* of a degree with a heating coil. And then there's the jeweled pivot, the point that goes into the jewel, like in a watch. So you see, it must be very expensive—I don't even know how expensive.

Student: Hasn't there been some work done on a sort of gyroscope that's a weight on the end of a flexible rod?

Feynman: Yes, yes. They have been trying to design other ways, other methods.

Student: Wouldn't that reduce the bearing problem?

Feynman: Well, it reduces one thing and creates something else.

Student: Is it being used?

Feynman: Not that I know of. The gyros we've discussed are the only ones that are actually used so far, and I don't think the others are in a position to match them *yet,* but they're close. It's a frontier subject. People are still designing new gyros, new devices, new ways, and it may well be that one of them will solve the problems, for instance, this insanity of having to have the axle bearings so accurate. If you play with the gyro for a while you will see that the friction on its *axle* is *not* small. The reason is, if the bearings were made too frictionless, the axle would wobble, and you'd have to worry about that tenth of a millionth of an inch—which is ridiculous. There must be a better way.

Student: I used to work in a machine shop.

Feynman: Then you can appreciate what is meant by a tenth of a millionth of an inch: it's impossible!

Other Student: What about ferroceramics?

Feynman: This business of supporting a superconductor in a magnetic field? Apparently if there's a fingerprint on the sphere, then the currents that are generated by the changing field make a little bit of loss. They're trying to get the thing straightened out, but it doesn't work *yet.*

There're a lot of other clever ideas, but I only wanted to show one in its final, engineered form, with all the details.

Student: The springs on that thing are *awfully* fine.

Feynman: Yeah. Not only are they fine, in the sense that they're small, but they're fine in the sense of the way they're made: you know, they're very good steel, spring steel, everything just right.

This kind of gyro is really impractical. It's *so* difficult to get it as accurate as it needs to be. It has to be made in rooms in which there's absolutely no dust—the people wear special coats, gloves, booties, and masks, because if there's one grain of dirt in one of these things, it makes the friction wrong. I'll bet they throw away more than they make successfully, because everything has to be *so* carefully built. It's not just some little thing you put together; it's quite difficult. This remarkable precision is just about at the edge of our present ability, so it's interesting, and any improvement you can invent or design into it, of course, would be a great thing.

One of the major problems is when the can's axis gets off center, and the thing turns; then you measure the twist around the wrong axis, and so you get a funny answer. But it seems to me self-evident (or almost—I may be wrong) that that's not *essential;* that there must be some way to support a rotating thing, so that the support follows the center of gravity. At the same time, you can measure that it's being twisted, because twisting is a different thing than having the center of gravity off.

What we would like to do is get a device that directly measures the twist about the center of gravity. If we could figure some way that the thing that's measuring the twist is *sure* to measure it about the center of gravity, it wouldn't make any difference if the center of gravity wobbled. If the whole platform always wobbled with just the *same* kind of motion as the thing you're trying to measure, then there'd be no way to get *out* of it. But this off-center wheel is *not* exactly the same as the thing you want to measure, so there must be some way out.

Student: In general, are mechanical/analog integrators on the way out, in favor of the electrical/digital ones?

Feynman: Well, yes.

Most integrating devices are electrical, but there are two general types. One is what they call "analog": such devices use a *physical method,* one for which the results of a measurement is an integral of something. For example, if you have a resistor and you develop a certain voltage, you get a certain current through the resistor, which is proportional to the voltage. But if you measure the total charge, not the current, that's the integral of the current, you see. When we integrated an acceleration by measuring an angle— that was a mechanical example. You can integrate in various ways of this kind, and it doesn't make any difference if it's mechanical or electrical— usually, it's electrical—but still, it's an analog method.

Then there's another way, and that's to get the signal out and to make the signal, for instance, into a frequency: the thing makes a lot of pulses, and when the signal is stronger, it makes pulses more rapidly. And then you *count* the pulses, you see?

Student: And integrate the number of pulses?

Feynman: Just *count* the pulses; you could count them on a device like one of those little pedometers, where you push it once for each pulse, or you could do the same thing electrically, with tubes flipping back and forth. Then, if you want to integrate that again, you can do something *numerically*—like we did our numerical integration on the blackboard. You can make essentially an adding machine—not an integrator, but an adding machine—and we use the adding machine to add the numbers together, and *those* numbers will have no appreciable errors in them if you design it right. So the errors due to the integrating devices can be reduced to zero, though the errors in the measuring equipment, from friction and so on, are still there.

They don't *use* digital integrators much in actual rockets and submarines—yet. But they're coming to that. They might as well get rid of the errors which are produced by the inaccuracies of the integrating machinery—and they *can* be gotten rid of, once you convert the signal to what they call digital information—dots—countable things.

Student: And then you just have a digital computer?

Feynman: Then you just have some kind of little digital computer that does two integrations, numerically. That's better than doing it the analog way in the long run.

Computing is mostly analog at the moment, but it's very likely that it'll turn into digital—in a year or two, probably—because that has no errors in it.

Student: You could use hundred-megacycle logic!

Feynman: It isn't the speed that's essential; it's simply a question of design. Analog integrators are getting so they're not quite accurate enough now, and so it's easiest to just change to digital. That's probably the next step, I would guess.

But the real problem, of course, is the gyro itself; *that* has to be made better and better.

Student: Thanks a lot for the lecture on applications. Do you think maybe later in the term you'll do more?

Feynman: You like that kind of stuff, about applications?

Student: I'm thinking about going into engineering.

Feynman: Okay. Well, this is one of the most beautiful things in mechanical engineering, of course.
   Let's try it . . .
   —Did it turn on?

Student: No. I guess it's not plugged in.

Feynman: Oh, excuse me. Here. I got it. *Now* switch it on.

Student: It says 'OFF' when I do that.

Feynman: *What?* I don't know what happened. Never mind. I'm sorry.

Another Student: Could you go over again how the Coriolis force works on a gyroscope?

Feynman: Yes.

Student: I can see how it works on the merry-go-round, already.

Feynman: All right. Here's a wheel which is turning on its axle—like a merry-go-round which is turning. I want to show that in order to *rotate the axle,* I have to *resist the precession* . . . or, that there'll be strains in the rods that support the axle, okay?

Student: Okay.

Feynman: Now, let's try to watch the way a particular particle of matter *on* the gyro wheel *actually* moves when we rotate the axle.

If the wheel *weren't* turning, the answer would be that the particle goes in a circle. There's centrifugal force on it, which is balanced by the strains on the spokes of the wheel. But the wheel *is* turning very rapidly. So when we rotate the axle, the piece of matter moves, and the wheel has also turned, you see? First it's here; now it's here: we've moved up to here, but the gyro turned. So the little piece of matter moves in a curve. Now, when you go around a curve, you've got to pull—it makes centrifugal force, if it's going in a curve. This force is not balanced by the spokes, which are radial; it must be balanced by some *sideways* push on the wheel.

Student: Oh! Yeah!

Feynman: So in order to *hold* this axle while it rotates I have to push sideways on it. You follow?

Student: Yeah.

Feynman: There's just one more point to make. You might ask, "If there's a sideways force, why doesn't the whole *gyro* move?" And the answer is, of course, that the *other side* of the wheel is moving the *opposite* way. And if you go through the same game, following a particle on the other side of the wheel when it's turning, it makes an opposite force on that side. So, there's no net force on the gyroscope.

Student: I'm starting to see it, but I can't see what difference the rotation of the wheel makes.

Feynman: Well, you see, it makes all the difference in the world. And the faster it goes around, the stronger is the effect—although it takes a little fiddling around to see why. Because if it goes faster, then the curve the particle makes isn't as sharp. On the other hand, it's going faster, and it's a problem of checking one against the other. Anyway, it turns out the force is greater when it's going faster—proportional to the speed, in fact.

Another Student: Dr. Feynman, . . .

Feynman: Yes, sir.

Student: Is it true that you can multiply seven-digit numbers in your head?

Feynman: No. It is *not true.* It's not even true that I can multiply *two*-digit numbers in my head. I can do *one*-digit numbers only.

Student: Do you know any philosophy teachers at Central College in Washington?

Feynman: Why?

Student: Well, I have a friend there. I hadn't seen him for a while and during Christmas vacation he asked me what I've been doing. I told him I was going to Caltech. So he asked, "Do you have a teacher there named Feynman?"—because his philosophy teacher told him that there's a guy named Feynman at Caltech who could multiply seven-digit numbers in his head.

Feynman: Not true. But I can do other things.

Student: Can I take some pictures of the apparatus?

Feynman: Sure! You want a close picture, or what?

Student: I think this'll do. But first, one to remember *you* by.

Feynman: I'll remember *you.*

5 *Selected Exercises*[1]

The following exercises are grouped into sections according to the chapters of *Exercises in Introductory Physics*. In parentheses the location of the corresponding subject matter in *The Feynman Lectures on Physics,* Volumes I–III, is provided. For example, the subject matter of the exercises in Section 5-1, "Conservation of energy, statics (Vol. I, Ch. 4)" is discussed in *The Feynman Lectures on Physics*, Volume I, Chapter 4.

Within each section the exercises are subdivided into categories according to degree of difficulty. In the order in which they appear in each section, these are: easy exercises (*), intermediate exercises (**), and more sophisticated and elaborate exercises (***). The average student should have little trouble solving the easy exercises, and should be able to solve most of the intermediate exercises within a reasonable time—perhaps ten to twenty minutes each. The more sophisticated exercises generally require a deeper physical insight or more extensive thought, and will be of interest principally to the better student.

5-1 Conservation of energy, statics (Vol. I, Ch. 4)

*1-1 A ball of radius 3.0 cm and weight 1.00 kg rests on a plane tilted at an angle α with the horizontal and also touches a vertical wall. Both surfaces have negligible friction. Find the force with which the ball presses on each plane.

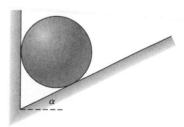

FIGURE 1-1

[1]From *Exercises in Introductory Physics*, by Robert B. Leighton and Rochus E. Vogt, 1969, Addison-Wesley, Library of Congress Catalog Card No. 73-82143. See **The exercises** in Michael Gottlieb's introduction, page xiv.

*1-2   The system shown is in static equilibrium. Use the principle of virtual work to find the weights A and B. Neglect the weight of the strings and the friction in the pulleys.

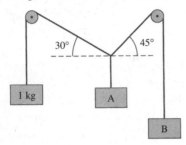

FIGURE 1-2

*1-3   What horizontal force F (applied at the axle) is required to push a wheel of weight W and radius R over a block of height h?

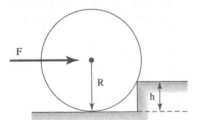

FIGURE 1-3

**1-4   A mass M_1 slides on a 45° inclined plane of height H as shown. It is connected by a flexible cord of negligible mass over a small pulley (neglect its mass) to an equal mass M_2 hanging vertically as shown. The length of the cord is such that the masses can be held at rest both at height H/2. The dimensions of the masses and the pulley are negligible

compared to H. At time t = 0 the two masses are released.

a) For t > 0 calculate the vertical acceleration of M_2.

b) Which mass will move downward? At what time t_1 will it strike the ground?

c) If the mass in (b) stops when it hits the ground, but the other mass keeps moving, show whether or not it will strike the pulley.

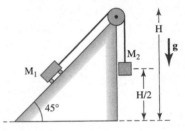

FIGURE 1-4

**1-5   A plank of weight W and length $\sqrt{3}R$ lies in a smooth circular trough of radius R. At one end of the plank is a weight W/2. Calculate the angle θ at which the plank lies when it is in equilibrium.

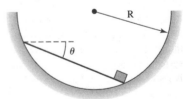

FIGURE 1-5

1-6 An ornament for a courtyard at a World's Fair is to be made up of four identical, frictionless metal spheres, each weighing $2\sqrt{6}$ ton-wts. The spheres are to be arranged as shown, with three resting on a horizontal surface and touching each other; the fourth is to rest freely on the other three. The bottom three are kept from separating by spot welds at the points of contact with each other. Allowing for a factor of safety of 3, how much tension must the spot welds withstand?

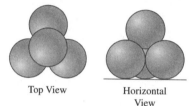

Top View   Horizontal View

FIGURE 1-6

1-7 A bobbin of mass $M = 3$ kg consists of a central cylinder of radius $r = 5$ cm and two end plates of radius $R = 6$ cm. It is placed on a slotted incline on which it will roll but not slip, and a mass $m = 4.5$ kg is suspended from a cord wound around the bobbin. It is observed that the system is in static equilibrium. What is the angle of tilt θ of the incline?

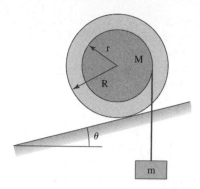

FIGURE 1-7

1-8 A cart on an inclined plane is balanced by the weight w. All parts have negligible friction. Find the weight W of the cart.

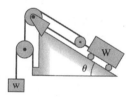

FIGURE 1-8

1-9 A tank of cross-sectional area A contains a liquid having density ρ. The liquid squirts freely from a small hole of area a at a distance H below the free surface of the liquid. If the liquid has no internal friction (viscosity), with what speed does it emerge?

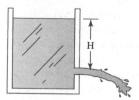

FIGURE 1-9

5-2 Kepler's laws and gravitation (Vol. I, Ch. 7)

*2-1 The eccentricity of the earth's orbit is 0.0167. Find the ratio of its maximum speed in its orbit to its minimum speed.

**2-2 A true "Syncom" (geosynchronous) satellite rotates synchronously with the earth. It always remains in a fixed position with respect to a point P on the earth's surface.

a) Consider the straight line connecting the center of the earth with the satellite. If P lies on the intersection of this line with the earth's surface, can P have any geographic latitude or what restrictions do exist? Explain.

b) What is the distance r_s from the earth's center of a Syncom satellite of mass m? Express r_s in units of the earth-moon distance r_{em}.

Note: Consider the earth a uniform sphere. You may use $T_m = 27$ days for the moon's period.

5-3 Kinematics (Vol. I, Ch. 8)

*3-1 A Skyhook balloon with a scientific payload rises at a rate of 1000 feet per minute. At an altitude of 30,000 feet the balloon bursts and the payload freefalls. (Such disasters do occur!)

a) For what length of time was the payload off the ground?

b) What was the payload's speed at impact?

Neglect air-drag.

*3-2 Consider a train that can accelerate with an acceleration of 20 cm s^{-2} and slow down with a deceleration of 100 cm s^{-2}. Find the minimum time for the train to travel between two stations 2 km apart.

*3-3 If you throw a small ball vertically upward in real air with drag, does it take longer to go up or come down?

**3-4 In a lecture demonstration a small steel ball bounces on a steel plate. On each bounce the downward speed of the ball arriving at the plate is reduced by a factor e in the rebound, i.e.,

$$v_{upward} = e \cdot v_{downward}.$$

If the ball was initially dropped from a height of 50 cm above the plate at time $t = 0$, and if 30 seconds later the silencing of a microphone sound indicated all bouncing had ceased, what was the value of e?

**3-5 The driver of a car is following a truck when he suddenly notices that a stone is caught between two of the rear tires of the truck. Being a safe driver (and a physicist too), he immediately increases his distance to the truck to 22.5 meters, so as not to be hit by the stone in case it comes loose. At what speed was the truck traveling? (Assume the stone does not bounce after hitting the ground.)

***3-6 A Caltech freshman, inexperienced with suburban traffic officers, has just received a ticket for speeding. Thereafter, when he comes upon one of the "Speedometer Test" sections on a level stretch of highway, he decides to check his speedometer reading. As he passes the "0" start of the marked section, he presses on his accelerator and for the entire period of the test he holds his car at constant acceleration. He notices that he passes the 0.10 mile post 16 seconds after

starting the test, and 8.0 seconds later he passes the 0.20 mile post.

a) What should his speedometer have read at the 0.20 mile post?

b) What was his acceleration?

***3-7 On the long horizontal test track at Edwards AFB, both rocket and jet motors can be tested. On a certain day, a rocket motor, started from rest, accelerated constantly until its fuel was exhausted, after which it ran at constant speed. It was observed that this exhaustion of the rocket fuel took place as the rocket passed the midpoint of the measured test distance. Then a jet motor was started from rest down the track, with a constant acceleration for the entire distance. It was observed that both rocket and jet motors covered the test distance in exactly the same time. What was the ratio of the acceleration of the jet motor to that of the rocket motor?

5-4   Newton's laws (Vol. I, Ch. 9)

*4-1 Two objects of mass m = 1 kg each, connected by a taut string of length L = 2 m, move in a circular orbit with constant speed V = 5 m s^{-1} about their common center C in a zero-g environment. What is the tension in the string in newtons?

FIGURE 4-1

**4-2 What horizontal force F must be constantly applied to M so that M_1 and M_2 do not move relative to M? Neglect friction.

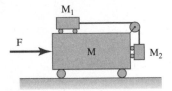

FIGURE 4-2

**4-3 An early arrangement for measuring the acceleration of gravity, called Atwood's Machine, is shown in the figure. The pulley P and cord C have negligible mass and friction. The system is balanced with equal masses M on each side as shown (solid line), and then a small rider m is added to one side. The combined masses accelerate through a certain distance h, the rider is caught on a ring, and the two equal masses then move on with constant speed, v. Find the value of g that corresponds to the measured values of m, M, h, and v.

***4-4 A painter weighing 180 lb working from a "bosun's" chair hung down the side of a tall building, desires to move in a hurry. He pulls down on the fall rope with such a force that he presses against the chair with only a force of 100 lb. The chair itself weighs 30.0 lb.

a) What is the acceleration of the painter and the chair?

b) What is the total force supported by the pulley?

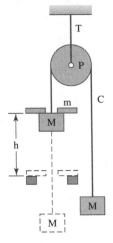

FIGURE 4-3

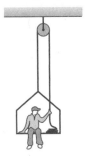

FIGURE 4-4

***4-5** A space traveler about to leave for the moon has a spring balance and a 1.0 kg mass A, which when hung on the balance on the earth gives the reading of 9.8 newtons. Arriving at the moon at a place where the acceleration of gravity is not known exactly but has a value of about 1/6 the acceleration of gravity at the earth's surface, he picks up a stone B which gives a reading of 9.8 newtons when weighed on the spring balance. He then hangs A and B over a pulley as shown in the figure and observes that B falls with an acceleration of 1.2 m s^{-2}. What is the mass of stone B?

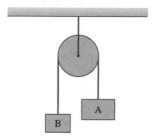

FIGURE 4-5

5-5 Conservation of momentum (Vol. I, Ch. 10)

***5-1** Two gliders are free to move on a horizontal air track. One is stationary and the other collides with it perfectly elastically. They rebound with equal and opposite velocities. What is the ratio of their masses?

***5-2** A machine gun mounted on the north end of a 10,000 kg, 5 m long platform, free to move on a horizontal air-bearing, fires bullets into a thick target mounted on the south end of the platform. The gun fires 10 bullets of mass 100 g each every second at a muzzle velocity of 500 m s^{-1}.

a) Does the platform move?

b) In which direction?

c) How fast?

****5-3** The end of a chain, of mass per unit length μ, at rest on a tabletop at t = 0, is lifted vertically at a constant speed v. Evaluate the upward lifting force as a function of time.

FIGURE 5-3

***5-4** The speed of a rifle bullet may be measured by means of a ballistic pendulum. The bullet, of known mass m and unknown speed V, embeds itself in a stationary wooden block of mass M, suspended as a pendulum of length L. This sets the block to swinging. The amplitude x of swing may be measured and, using conservation of energy, the velocity of the block immediately after impact may be found. Derive an expression for the speed of the bullet in terms of m, M, L, and x.

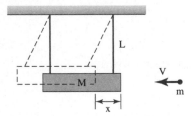

FIGURE 5-4

***5-5** Two equally massive gliders, moving on a level air track at equal and opposite velocities, v and −v, collide almost elastically, and rebound with slightly smaller speeds. They lose a fraction f ≪ 1 of their kinetic energy in the collision. If these same gliders collide with one of them initially at rest, with what speed will the second glider move after the collision? (This small residual speed Δv may easily be measured in terms of the final speed v of the originally stationary glider, and thus the elasticity of the spring bumpers may be determined.)

Note: If $x \ll 1$, then
$$\sqrt{1 - x} \approx 1 - \tfrac{1}{2}x.$$

***5-6** An earth satellite of mass 10 kg and average cross-sectional area 0.50 m^2 is moving in a circular orbit at 200 km altitude where the molecular mean free paths are many meters and the air density is about 1.6×10^{-10} kg m^{-3}. Under the crude assumption that the molecular impacts with the satellite are effectively inelastic (but that the molecules do not literally stick to the satellite but drop away from it at low relative velocity), calculate the retarding force that the satellite would experience due to air friction. How should such a frictional force vary with velocity? Would the satellite's speed decrease as a result of the net force on it? (Check the speed of a circular satellite orbit versus height.)

5-6 Vectors (Vol. I, Ch. 11)

6-1 A man standing on the bank of a river 1.0 mi wide wishes to get to a point directly opposite him on the other bank. He can do this in two ways: (1) head somewhat upstream so that his resultant motion is straight across, (2) head toward the opposite bank and then walk up along the bank from the point downstream to which the current has carried him. If he can swim 2.5 mi hr^{-1} and walk 4.0 mi hr^{-1}, and if the current is 2.0 mi hr^{-1}, which is the faster way to cross, and by how much?

6-2 A motorboat that runs at a constant speed V relative to the water is operated in a straight river channel where the water is flowing smoothly with a constant speed R. The boat is first sent on a round trip from its anchor point to a point a distance d directly upstream. It is then sent on a round trip from its anchor point to a point a distance d away directly across the stream. For simplicity assume that the boat runs the entire distance in each case at full speed and that no time is lost in reversing course at the end of the outward lap. If t_V is the time the boat took to make the round trip in line with the stream flow, t_A the time the boat took to make the round trip across the stream, and t_L the time the boat would take to go a distance 2d on a lake.

a) What is the ratio t_V/t_A?

b) What is the ratio t_A/t_L?

6-3 A mass m is suspended from a frictionless pivot at the end of a string of arbitrary length, and is set to whirling in a horizontal circular path whose plane is a distance H below the pivot point. Find the period of revolution of the mass in its orbit.

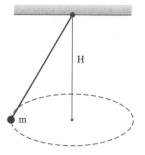

FIGURE 6-3

6-4 You are on a ship traveling steadily east at 15 knots. A ship on a steady course whose speed is known to be 26 knots is observed 6.0 mi due south of you; it is later observed to pass behind you, its distance of closest approach being 3.0 mi.

a) What was the course of the other ship?

b) What was the time between its position south of you and its position of closest approach?

5-7 Nonrelativistic two-body collisions in 3 dimensions (Vol. I, Chs.10 and 11)

**7-1 A moving particle of mass M collides perfectly elastically with a stationary particle of mass m < M. Find the maximum possible angle through which the incident particle can be deflected.

**7-2 An object of mass m_1, moving with a linear speed v in a laboratory system, collides with an object of mass m_2 which is at rest in the laboratory. After the collision, it is observed that $(1 - \alpha^2)$ of the kinetic energy in the CM system was lost in the collision. What was the percentage loss of energy in the *laboratory* system?

**7-3 A proton with kinetic energy 1 MeV collides elastically with a stationary nucleus and is deflected through 90°. If the proton's energy is now 0.80 MeV, what was the mass of the target nucleus in units of the proton mass?

5-8 Forces (Vol. I, Ch. 12)

*8-1 Two masses, $m_1 = 4$ kg and $m_3 = 2$ kg, are connected with cords of negligible weight over essentially frictionless pulleys to a third mass, $m_2 = 2$ kg. The mass m_2 moves on a long table with a coefficient of friction

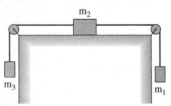

FIGURE 8-1

$\mu = 1/2$. What is the acceleration of mass m_1 after the system is released from rest?

**8-2 A 5 g bullet is fired horizontally into a 3 kg wooden block resting on a horizontal surface. The coefficient of sliding friction between the block and surface is 0.2. The bullet remains embedded in the block, which is observed to slide 25 cm along the surface. What was the velocity of the bullet?

8-3 In their investigation at the scene of an automobile accident, the police found, by measurement, that car A left skid marks 150 feet long before it collided with car B. It was also known that the coefficient of friction between rubber and the pavement at the scene of the accident was not less than 0.6. Show that car A must have been exceeding the posted speed limit of 45 mph just prior to the accident. (Note that 60 mph = 88 feet/sec and acceleration due to gravity = 32 feet/sec^2).

8-4 An air-conditioned school bus is approaching a railway crossing. One of the children has tied a hydrogen filled balloon to a seat. You observe that the anchor line of the balloon makes an angle of 30° with the vertical in the direction of motion. Is the driver decelerating or accelerating the bus, and by how much? (Would a highway patrol officer commend the driver for his skill?)

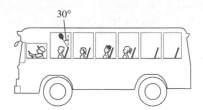

FIGURE 8-4

***8-5** A particle of weight W rests on a rough inclined plane that makes an angle α with the horizontal.

a) If the coefficient of static friction $\mu = 2 \tan \alpha$, find the least *horizontal* force H_{min}, acting transverse to the slope of the plane that will cause the particle to move.

b) In what direction will it go?

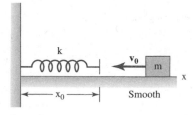

FIGURE 8-5

5-9 Potentials and Fields (Vol. I, Chs. 13 and 14)

*9-1 A mass m collides with a spring of spring constant k. At what point does it first come to rest? Neglect the mass of the spring.

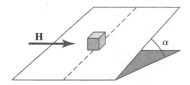

FIGURE 9-1

*9-2 A hollow spherical asteroid travels freely through space. There is a small particle of mass m in its interior. At what point in the interior will the particle be in equilibrium position?

*9-3 The speed needed for a body to leave the earth's gravitational field is (approximately) 7.0 mi s^{-1}. If an interplanetary probe is given an initial speed of 8.0 mi s^{-1} just above the earth's atmosphere, with what speed relative to the earth will it be traveling when it is at a distance of 10^6 mi from the earth?

**9-4 A small, frictionless car coasts on an inclined track with a circular loop-the-loop of radius R at its lower end. From what height H above the top of the loop must the car start in order to traverse the loop without leaving the track?

**9-5 A flexible cable of length L that weighs M kg m^{-1} hangs over a pulley of negligible mass, radius, and friction. Initially, the cable is just balanced. It is given a slight push to unbalance it, and it proceeds to accelerate. Find its speed as the end flies off the pulley.

**9-6 A particle starts from rest at the top of a frictionless sphere of radius R and slides on the sphere under the force of gravity. How far below its starting point does it get before flying off the sphere?

**9-7 An automobile weighing 1,000 kg is powered by an engine whose rated power is 120 kW. If the engine develops this power at a speed of 60 km h^{-1}, what is the maximum acceleration the car can have at this speed?

**9-8 World records (1960) for the shotput, the discus, and the javelin were respectively 19.30 m, 59.87 m, and 86.09 m. The masses of the missiles involved are respectively 7.25 kg, 2 kg, and 0.8 kg. Compare the work done by each champion in making his record toss, assuming that each trajectory starts at an elevation of 1.80 m above level ground and has an initial elevation of 45°. Neglect air resistance.

***9-9 A satellite of mass m moves in a circular orbit around an asteroid of mass M $(M \gg m)$. If the asteroid's mass was suddenly[2] reduced to one-half its former value, what would happen to the satellite? Describe its new orbit.

---

[2]How it could happen: The satellite is placed in orbit at a large distance from the asteroid to monitor the test of a nuclear device on the asteroid. The explosion expels half the asteroid's mass without directly affecting the distant satellite.

5-10   Units and dimensions (Vol. I, Ch. 5)

*10-1   Moe and Joe, two cosmic physicists who grew up on different planets, meet at an interplanetary symposium on weights and measures to discuss the establishment of a universal system of units. Moe proudly describes the merits of the MKSA system, used in every civilized region of the earth. Joe equally proudly describes the beauties of the $M'K'S'A'$ system, used everywhere else in the solar system. If the constant factors relating the basic mass, length, and time standards of the two systems are μ, λ, and τ, such that

$$m' = \mu m, \quad l' = \lambda l, \quad \text{and} \quad t' = \tau t$$

what factors are needed to convert the units of velocity, acceleration, force, and energy between the two systems?

**10-2   If a scale model of the solar system is made, using materials of the same respective average densities as the sun and planets, but reducing all linear dimensions by a scaling factor k, how will the periods of revolution of the planets depend on k?

5-11   Relativistic energy and momentum (Vol. I, Chs. 16 and 17)

*11-1

a) Express the momentum of a particle in terms of its kinetic energy T and rest energy $m_0 c^2$.

b) What is the speed of a particle whose kinetic energy is equal to its rest energy?

**11-2   A pion ($m_\pi = 273\ m_e$) at rest decays into a muon ($m_\mu = 207\ m_e$) and a neutrino ($m_\nu = 0$). Find the kinetic energy and momentum of the muon and the neutrino in MeV.

**11-3   A particle of mass m, moving at speed $v = 4c/5$, collides inelastically with a similar particle at rest.

a) What is the speed of the composite particle?

b) What is its mass?

**11-4   A proton-antiproton pair may be created in the absorption of a photon (γ) by a proton at rest.

$$\gamma + P \rightarrow P + (P + \bar{P})$$

What minimum energy E_γ must the photon have? (Express E_γ in terms of proton rest energy $m_p c^2$).

5-12 Rotations in two dimensions, the center mass (Vol. I, Chs. 18 and 19)

**12-1 A disc of uniform density has a hole cut out of it, as shown. Find the center of mass.

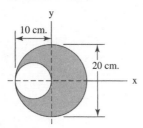

y

10 cm.

20 cm.

x

FIGURE 12-1

**12-2 A solid cylinder has a density that varies by quadrants as shown, with the numbers indicating relative densities. If the x-y axes are as indicated, what is the equation of the line drawn through the origin and through the center of mass?

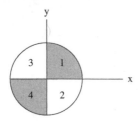

y

3 | 1

x

4 | 2

FIGURE 12-2

**12-3 From a square piece of uniform sheet metal an isosceles triangle is to be cut out from one edge, as shown, such that the remaining metal, when suspended from the apex P of the cut, will remain in equilibrium in any position. What is the altitude of the cutout triangle?

a

P

FIGURE 12-3

**12-4 Masses M_1 and M_2 are placed at the opposite ends of a rigid rod of length L and negligible mass; the dimensions of M_1 and M_2 are negligible compared to L. The rod is to be set rotating about an axis perpendicular to it. Through what point on this rod should this axis pass in order that the work required to set the rod rotating with an angular speed ω_0 shall be a minimum?

***12-5 A uniform brick of length L is laid on a smooth horizontal surface. Other equal bricks are now piled on as shown, so that the sides form continuous planes, but the ends are offset at each brick from the previous brick by a distance L/a, where a is an integer. How many bricks can be used in this manner before the pile topples over?

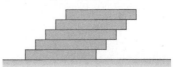

FIGURE 12-5

***12-6** A rotating governor, as shown, is to be designed to shut off power when the machine to which the governor is directly connected reaches a speed of 120 rpm. The operating collar C weighs 10.0 lb and slides without friction on the vertical shaft AB. C is so designed to shut off power when the distance AC reduces to 1.41 ft. If the four links of the governor framework are each 1.00 ft long between frictionless pivots and are relatively massless, what value should the masses M have so that the governor will operate as planned?

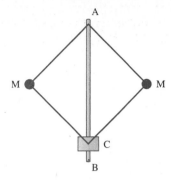

FIGURE 12-6

5-13 Angular momentum, the moment of inertia (Vol. I, Chs. 18 and 19)

*13-1** A straight, uniform wire of length L and mass M is bent at its midpoint to form the angle θ. What is its moment of inertia for an axis passing through the point A, perpendicular to the plane determined by the bent wire?

FIGURE 13-1

*13-2** A mass m is hung from a string wound around a solid circular cylinder of mass M and radius r, pivoted on bearings of negligible friction as shown. Find the acceleration of m.

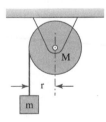

FIGURE 13-2

**13-3 A horizontal thin rod of mass M, length L rests at one end on a support and is suspended by a string at the other end. What force is exerted by the rod on the support immediately after the string is burned?

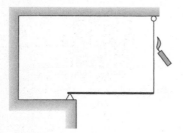

FIGURE 13-3

**13-4 Starting from rest, a symmetrical object rolls (without slipping) down an incline of height h. The moment of inertia of the object about its center of mass is I, the mass is M, and the radius of the rolling surface in contact with the incline is r. Determine the linear velocity of the center of mass at the bottom of the incline.

**13-5 On an endless belt that is inclined at an angle θ with the horizontal, a uniform cylinder is placed, its axis horizontal and perpendicular to the edge of the belt.

The surfaces are such that the cylinder can roll without slipping on the belt. How should the belt be caused to move so that, when released, the axis of the cylinder does not move?

**13-6 The hoop H of radius r rolls without slipping down the incline. The starting height h is such that the hoop acquires a velocity just sufficient to "loop the loop"—i.e., the hoop just maintains contact with the circular track at point P. What is h?

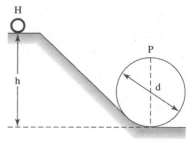

FIGURE 13-6

***13-7 A uniform bowling ball of radius R and mass M is initially launched so that it is sliding with speed V_0 without rolling on an alley with a coefficient of friction μ. How far does the ball go before it starts rolling without slipping, and what is its speed then?

***13-8** An amusing trick is to press a finger down on a marble, on a horizontal table top, in such a way that the marble is projected along the table with an initial linear speed V_0 and an initial backward rotational speed ω_0, ω_0 being about a horizontal axis perpendicular to V_0. The coefficient of sliding friction between marble and table top is constant. The marble has radius R.

a) What relationship must hold between V_0, R, and ω_0 for the marble to slide to a complete stop?

b) What relationship must hold between V_0, R, and ω_0 for the marble to skid to a stop and then start returning toward its initial position, with a final *constant* linear speed of $3/7\,V_0$?

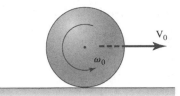

FIGURE 13-8

5-14 Rotation in three dimensions (Vol. I, Ch. 20)

*14-1** A jet airplane in which all the engines rotate in the direction of a right-handed screw advancing in the flight direction is executing a left turn. Does the gyroscopic effect of the engines tend to cause the airplane to:

a) roll right

b) roll left

c) yaw right

d) yaw left

e) pitch up

f) pitch down

14-2 Two equal masses are connected by a flexible string. An experimenter holds one mass in his hand and causes the other mass to whirl in a horizontal circle about the held mass; he then releases the held mass.

a) If the string breaks during the experiment, did it break before or after he released the masses?

b) If the string does not break, describe the motion of the masses subsequent to their release.

14-3 A thin circular wooden hoop of mass m and radius R rests on a horizontal frictionless plane. A bullet, also of mass m, moving with horizontal velocity v, strikes the hoop and becomes embedded in it as shown in the figure. Calculate the center-of-mass velocity, the angular momentum of the system about the CM, the angular velocity ω of the hoop, and the kinetic energy of the system, before and after collision.

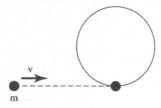

FIGURE 14-3

14-4 A thin rod of mass M and length L rests on a horizontal frictionless surface. A small piece of putty, also of mass M, and with velocity v directed perpendicularly to the rod, strikes one end and sticks, making an inelastic collision of very short duration.

a) What is the velocity of the center of mass of the system before and after the collision?

b) What is the angular momentum of the system about its center of mass just before the collision?

c) What is the angular velocity (about the center of mass) just after the collision?

d) How much kinetic energy is lost in the collision?

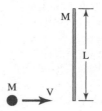

FIGURE 14-4

14-5 A thin uniform rod AB of mass M and length L is free to rotate in a vertical plane about a horizontal axle at end A. A piece of putty, also of mass M, is thrown with velocity V horizontally at the lower end B while the bar is at rest. The putty sticks to the bar. What is the minimum velocity of the putty before impact that will make the bar rotate *all the way around A*?

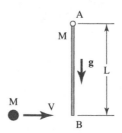

FIGURE 14-5

**14-6 A turntable T_1 at rest has mounted on it a turntable T_2 rotating with angular velocity ω. At a certain time an internal clutch acts on the axle of T_2 to stop it with respect to T_1, but T_1 is free to revolve. T_1 alone has mass M_1 and moment of inertia I_1 about an axis A_1 through its center perpendicular to its plane; and T_2 has mass M_2 and I_2 about a similarly situated axis A_2; the distance between A_1 and A_2 is r. Find Ω for T_1 after T_2 stops. (Ω is the angular velocity of T_1.)

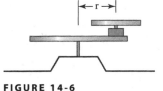

FIGURE 14-6

***14-7 An upright rod of mass M and length L is given an impulse J at its base, directed at 45° upward from the horizontal, which sends the rod flying. What value(s) should J have so that the rod lands vertically again (i.e., upright on the end at which J was applied)?

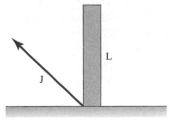

FIGURE 14-7

***14-8 A turntable of moment of inertia I_0 rotates freely on a hollow vertical axis. A cart of mass m runs without friction on a straight radial track on the turntable. A cord attached to the cart passes over a small pulley and then downward through the hollow axis. Initially the entire system is rotating at angular speed ω_0, and the cart is at a fixed radius R from the axis. The cart is then pulled inward by applying an excess force to the cord, and eventually arrives at radius r, where it is allowed to remain.

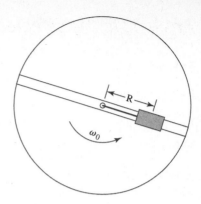

FIGURE 14-8

a) What is the new angular velocity of the system?

b) Show in detail that the difference in the energy of the system between the two conditions is equal to the work done by the centripetal force.

c) If the cord is released, with what radial speed dr/dt will the cart pass the radius R?

***14-9 A flywheel having the shape of a uniform thin circular plate of mass 10.0 kg and radius 1.00 m is mounted on a shaft passing through its CM but making an angle of 1°0' with its plane. If it rotates about this axis with angular velocity 25.0 radians s^{-1}, what torque must be supplied by the bearings?

Answers to the Selected Exercises

1-1

$$F_P = \frac{1}{\cos \alpha} \text{ kg-wt}$$

$$F_W = \tan \alpha \text{ kg-wt}$$

1-2

$$A = \left(\frac{1}{2} + \frac{\sqrt{3}}{2}\right) \text{ kg-wts}$$

$$B = \sqrt{\frac{3}{2}} \text{ kg-wts}$$

1-3

$$F = W\frac{\sqrt{h(2R-h)}}{R-h}$$

1-4

a) $a = -\frac{1}{2}\left(1 - \frac{1}{\sqrt{2}}\right)g$

b) $M_2, t_1 = \sqrt{\dfrac{2H}{g\left(1 - \dfrac{1}{\sqrt{2}}\right)}}$

c) No

1-5

$\theta = 30°$

1-6

2 ton-wts

1-7

$\theta = 30°$

1-8

$$W = \frac{4w}{\sin \theta}$$

1-9

$v = \sqrt{2gH}$

2-1

1.033

2-2

a) $\lambda = 0$

b) $r_s = \frac{1}{9}r_{em}$

3-1

a) $t = 1843.8$ s

b) $v \approx 1385 \text{ ft s}^{-1}$

3-2

≈ 155 s

3-3

down

3-4

$e \approx 0.98$

3-5

14.8 m s^{-1}

3-6

a) 52.5 mi hr^{-1}

b) 2.75 ft s^{-2}

3-7

$$a_J = \frac{8}{9}a_R$$

4-1

$T = 25$ N

4-2

$$F = \frac{M_2}{M_1}(M + M_1 + M_2)g$$

4-3

$$g = \frac{v^2(2M + m)}{2mh}$$

4-4

a) $a_{up} = g/3$

b) 280 lb

4-5

$m_B \approx 5.8$ kg

5-1

$m_2/m_1 = 3$

5-2

a) Yes

b) To the N

c) $V = 5 \times 10^{-4}$ m s^{-1}

5-3

$F = \mu v(v + gt)$

5-4

$$V = x\frac{m + M}{m}\sqrt{\frac{g}{L}}$$

5-5

$$\Delta v \approx v\frac{f}{4}$$

5-6

$F_R = 5.1 \times 10^{-3}$ N

$F_R \propto -v^2$

6-1

Method 2, by 4.0 min.

6-2

$$\frac{t_V}{t_A} = \frac{V}{\sqrt{V^2 - R^2}}$$

$$\frac{t_A}{t_L} = \frac{t_V}{t_A}$$

6-3

$$T = 2\pi\sqrt{\frac{H}{g}}$$

6-4

a) due N

b) 0.17 hr

7-1

$$\theta_{max} = \sin^{-1}\frac{m}{M}$$

7-2

$$\frac{\Delta T}{T}\bigg|_{lab} = \frac{(1 - \alpha^2)m_2}{m_1 + m_2}$$

7-3

$$\frac{M}{m_P} = 9$$

8-1

$$a = -\frac{g}{8}$$

8-2

$v_0 = 595$ m s^{-1}

8-3

51.8 mph

8-4

Accelerating

$$a = \frac{g}{\sqrt{3}} \text{ m s}^{-2}$$

8-5

a) $\sqrt{3}W \sin \alpha$

b) $\phi = 60°$

9-1

$$x_0 - x = x_0 - v_0\sqrt{\frac{m}{k}}$$

9-2

Anywhere

9-3

$v_\infty \approx 3.9$ mi s^{-1}

9-4

$H = \dfrac{1}{2}R$

9-5

$v = \sqrt{\dfrac{gL}{2}}$

9-6

$\dfrac{R}{3}$

9-7

7.2 m s^{-2}

9-8

≈ 625 J

≈ 570 J

≈ 330 J

9-9

The satellite would escape on a parabolic orbit.

10-1

$v' = \dfrac{\lambda}{\tau}v$

$a' = \dfrac{\lambda}{\tau^2}a$

$F' = \dfrac{\mu\lambda}{\tau^2}F$

$E' = \dfrac{\mu\lambda^2}{\tau^2}E$

10-2

T is independent of k.

11-1

a) $pc = T\left(1 + \dfrac{2m_0c^2}{T}\right)^{1/2}$

b) $\dfrac{v}{c} = \dfrac{\sqrt{3}}{2}$

11-2

$T_\mu = 4.1$ MeV

$T_\nu = 29.7$ MeV

$p_\mu = p_\nu = 29.7$ MeV/c

11-3

a) $c/2$

b) $\dfrac{4}{\sqrt{3}}m$

11-4

$E_\gamma = 4m_pc^2$ (3.8 GeV)

12-1

$x = 1.7$ cm

12-2

$y = \dfrac{1}{2}x$

12-3

$h = \dfrac{a}{2}(3 - \sqrt{3})$

12-4

$x = \dfrac{m_1L}{m_1 + m_2}$ (from m_2)

12-5

$n = a$

12-6

$M = 4.0$ lb

13-1

$$I = \frac{mL^2}{12}$$

13-2

$$a = \frac{mg}{m + \dfrac{M}{2}}$$

13-3

$$F = \frac{Mg}{4}$$

13-4

$$V_0 = r\sqrt{\frac{2Mgh}{I + Mr^2}}$$

13-5

$$a = 2g \sin \theta$$

13-6

$$h = \frac{3d}{2} - 3r$$

13-7

$$D = \frac{12V_0^2}{49\mu g}$$

$$V = \frac{5}{7}V_0$$

13-8

a) $V_0 = \dfrac{2}{5}R\omega_0$

b) $V_0 = \dfrac{1}{4}R\omega_0$

14-1
(e)

14-2
a) before

b) $V_{CM} = \dfrac{\ell}{2}\omega_0 \quad \omega = \omega_0$

(where ℓ is the length of the string)

14-3

$$V_{CM} = \frac{v}{2}$$

$$L = \frac{mvR}{2}$$

$$\omega = \frac{v}{3R}$$

$$\text{K.E.}\Big|_1 = \frac{mv^2}{2}$$

$$\text{K.E.}\Big|_2 = \frac{mv^2}{3}$$

14-4

a) $\dfrac{v}{2}$

b) $Mv\dfrac{L}{4}$

c) $\dfrac{6}{5}\dfrac{v}{L}$

d) 20%

14-5

$$V = \sqrt{8g\,L}$$

14-6

$$\Omega = \frac{I_2}{I_1 + I_2 + M_2r^2}\omega$$

14-7

$$J = M\sqrt{\frac{\pi gLn}{3}} \quad (n = \text{integer})$$

14-8

a) $\omega = \dfrac{I_0 + mR^2}{I_0 + mr^2}\omega_0$

b) (No answer was given.)

c) $v = \omega_0\sqrt{\dfrac{I_0 + mR^2}{I_0 + mr^2}(R^2 - r^2)}$

14-9

$T \sim 27\,\text{N m}$

Photo Credits

Index